HERITAGE COVENANTS & PRESERVATION
the Calgary Civic Trust

HERITAGE COVENANTS & PRESERVATION
the Calgary Civic Trust

UNIVERSITY OF
CALGARY
PRESS

Edited and Introduction by
Michael McMordie and Frits Pannekoek

Editorial Team:
Anne English, Kimberly E. Haskell,
and Sally Jennings

©2004 The Calgary Civic Trust
Published by the
University of Calgary Press
2500 University Drive NW
Calgary, Alberta, Canada T2N 1N4
www.uofcpress.com

No part of this publication may be reproduced, stored in a retrieval system or transmitted, in any form or by any means, without the prior written consent of the publisher or a licence from The Canadian Copyright Licensing Agency (Access Copyright). For an Access Copyright licence, visit www.accesscopyright.ca or call toll free to 1-800-893-5777.

We acknowledge the financial support of the Government of Canada through the Book Publishing Industry Development Program (BPIDP), the Alberta Foundation for the Arts and the Alberta Lottery Fund—Community Initiatives Program for our publishing activities. We acknowledge the support of the Canada Council for the Arts for our publishing program.

∞ This book is printed on acid-free paper. Printed and bound in Canada by AGMV Marquis. Cover design, Mieka West
Page design, typesetting and production by Samuel Smith Esseh.

Photo credits: Unless otherwise credited, Calgary photos courtesy Alberta Community Development, Heritage Resource Management Branch, Protection and Stewardship Section, and other Alberta photos courtesy Les Hurt.

Library and Archives Canada Cataloguing in Publication

  Heritage covenants and preservation : the Calgary Civic Trust / Edited and introduction by Michael McMordie and Frits Pannekoek ; editorial team, E. Anne English, Kimberly E. Haskell and Sally Jennings.

Proceedings of a workshop held at University of Calgary, Sept. 15, 2000.

Parks and heritage series, ISSN 1494-0426;9)
Includes bibliographical references and index.
ISBN 1-55238-133-1

  1. Historic preservation—Alberta—Calgary. I. McMordie, Michael II. Pannekoek, Frits, 1949-

NA109.C3H47 2004 363.6'9'0971338 C2004-905043-5

CONTENTS

Editors' Acknowledgments vii

Covenants: An Introduction 1

Workshop: Introduction 15

Section 1 — **The International Context 17**

1: The British Experience 19
Tim Butler, Solicitor to the National Trust, United Kingdom

2: The American Experience 35
Paul Edmondson, Vice President & General Council, National Trust for Historic Preservation

Section 2 — **Canadian Precedents 55**

3: Ontario Precedents 57
Jeremy Collins, Coordinator, Acquisition & Dispositions, Ontario Heritage Foundation

4: The Preservation and Planning Context in Alberta 77
Larry Pearson, Protection & Stewardship, Heritage Resources Management

5: The Calgary Context 95
Rob Graham, Heritage Planner, City of Calgary

Section 3 — **Financial and Legal Considerations 119**

6: Tax Implications: Financial and Legal Considerations 121
Marc Denhez, Heritage Lawyer

7: Heritage Canada Policy: The Canadian Tax Context 133
Doug Franklin, Heritage Canada Foundation

8: Calculating the Market Value: Appraisal Methods 141
Jason Ness, Calgary Civic trust

Section 4 **Discussion 147**

9: A Review of Key Issues and Design of a Covenant 149
Michael McMordie

10: Realities of the Development Industry: The Impact of Covenanting on Development- A Panel Discussion 175

11: Stewardship through Covenanting: Moving Forward 199

Biographies: Speakers and Panel Participants 205

Appendices 214

Bibliography 240

Notes 242

Index 245

Editors' Acknowledgments

Books like this one can only be achieved with dedication from a number of people. The whole of the Trust were involved in some way. We wish to acknowledge the generous support received from The Alberta Law Foundation and the Alberta Historical Resources Foundation. We would also like to express our gratitude for the support drawn from the University of Calgary. We must also acknowledge with gratitude the co-operation and assistance of Heritage Canada, whose annual conference took place immediately following this workshop. That co-operation made it possible to bring a number of presenters, who were vital to the discussion on the subject of covenanting, to the workshop.

We particularly acknowledge and thank Sally Jennings, Executive Director of the Civic Trust, for her work in organizing the workshop and to the volunteers and participants for making the workshop a success. This book could not have been produced without the strong team of recorders, editors and bibliographers: E. Anne English, Kimberly E. Haskell, and Sally Jennings. Putting the complex set of recordings into order and into a readable format required considerable dedication. We decided early on that rather than convert the "conversations" into academic essays we would leave them as close to the original as possible, only eliminating repetition where it occurred.

We would also like to thank Ann Cowie and Becky Naiden for the work they have done during the conference and afterwards to ensure that all was coordinated.

Dr. Michael McMordie, President
Dr. Frits Pannekoek, Secretary

Covenants: An Introduction

Frits Pannekoek and Michael McMordie

Until now in Alberta, provincial and municipal designations undertaken pursuant to the *Alberta Historical Resources Act* (R.S.A. 2000, c. H-9), first passed in 1970 as the *Alberta Heritage Act*, have been the most powerful tools for the preservation of historic properties in the province. Although the Act has been adjusted over time, it remains essentially unrevised. Designation (Sections 19, 20, and 26) involves the placement by government of a restriction on title. Generally these restrictions dictate that a designated property owner must give notice or secure approval from the designating authority (the Minister or the Municipality) to make any physical changes. There are several categories of designation. Provincially designated sites are truly of provincial significance and can only be altered with the approval of the Minister. Registered sites only need inform the Minister of their intention to alter the building or site. Municipal designations, which are by bylaw, require municipal approval. Currently there are several hundred buildings designated by the Province, but only several dozen by Municipalities, and the majority of these in Edmonton. The key issue in Alberta has been that of compensation. Under Section 50 of the Act, the Provincial government *may* pay compensation to a designated owner pursuant to regulations. These regulations have not as yet been crafted. Municipal governments *must* pay compensation to protesting designated site owners. This provision has put municipalities in a bind, although some like Edmonton have crafted imaginative solutions to solve the issue. Others have been less inclined to do so and tend to rely more on provincial powers to achieve their objectives.

The difficulties of preservation in Canada generally stem from the *Constitution Act* (formerly the *British North America Act*) of Canada, which has designated "Property and Civil Rights" (clause 92(13)) a provincial responsibility. Occasionally, the federal government will preserve those buildings under its jurisdiction, e.g., railway buildings or defence lands through statute, e.g., *Heritage Railway Stations Protection Act*, proclaimed in 1990,

Medalta Potteries, Medicine Hat, Alberta

or Treasury Board Policy, e.g., the Federal Heritage Building Review Office. However, except for these particular situations, federal designation through the National Historic Sites and Monuments Board never involves legal protection unless the Crown owns the resource or there are separate contribution agreements. These agreements are an appropriate solution, but again rely on direct federal subsidy and involve not-for-profits rather than individual citizen-owned buildings. The best example in Alberta is the outstanding work with Medalta Potteries – one of the key clay industries in Canada, at one time responsible for the manufacture of Canadian railroad and hotel dinnerware.

The Civic Trust believes that provincially based not-for-profits could become involved in the preservation of heritage structures, not to the exclusion of private development, federal agreements, or provincial or municipal designation, but rather as an alternative. Private trusts could either own property or, more important, own an interest in a property. In 1974 Alberta's Old Strathcona Foundation was established, and two years later it received an initial funding of $1 million from Heritage Canada, the provincial government, and the Devonian Foundations. For the most part, the Old Strathcona Foundation was successful, though it was never more than a transitory property owner and never protected its interests through covenants. Heritage Canada, on the other hand, did and continues to own (although it does little for enforce them) covenants on several Strathcona historic residences. However, given operating and maintenance costs, they were reluctant to become Canada's equivalent of the British or American national trusts. So Canada has no real national independent not-for-profit interested in and committed to the acquisition of interests in real property in a systematic manner. The closest is the Ontario Heritage Foundation, an arm of the Government of the Province of Ontario.

Covenanting in Canada is a somewhat complex issue as the proceedings of the conference illustrate. "Covenanting," regardless of jurisdiction, involves the donation or sale of the "heritage interest" in a building or landscape to a not-for-profit entity like the Calgary Civic Trust or the Crown. That interest is registered against the title of the property and is called a "covenant." The holder of the covenant has an obligation to monitor the agreement or covenant and ensure that its interests are maintained. Using this legal instrument, the variety of buildings preserved could be much greater and the decision to preserve would be motivated by the owner, the community, and the government.

Covenanting has been used successfully by many organization involved in the conservation field for protecting both buildings and natural resources. The Ontario Heritage Foundation (which holds 170 covenants), the National Trust in England, and the National Trust for Historic Preservation in the United States are but a few of the well-known not-for-profits that focus on covenants as their key heritage preservation instrument.

However, after consulting lawyers and professionals in the heritage preservation field in Calgary, the Calgary Civic Trust found there was little knowledge of the covenanting mechanism in Canada. In order to use this mechanism successfully, and to increase its use in Canada and particularly in Alberta, the Civic Trust held a two-day covenanting workshop on September 13–14, 2000. The Trust brought together experts in covenanting, lawyers, heritage professionals, community association members, potential covenanters, and the general public to learn from and exchange their experiences. The goals of the workshop were to educate Albertans on the principles of using covenants for heritage preservation, to establish covenanting in Alberta, to enable property owners to take actions to preserve their own heritage, and to develop covenanting guidelines and legal documentation for those interested in covenants.

This book is the outcome of the conference and has the same single focus: the potentials and problems that Canadian jurisdictions face in dealing with covenants. Prior to providing a context to covenants in Alberta, it will be useful to define the term "covenant" in detail, but in layman's terms. For this discussion, the terms "covenant" and its alternative "easement" are interchangeable: we have chosen to use "covenant."

Even though our legal system indicates that all land ultimately belongs to the Crown and that rights can be limited by the government, in Canada, real property is still considered in modern folklore to be a sacrosanct unit of land owned by an individual, shrouded in the myths of the "sanctity" of "the home as castle." What is not really understood is that "real property" rights are indeed plural. Simply put, they consist of individual and Crown rights, or a bundle of rights, which are mitigated by various considerations and agreements. For example, you might own land, but not its minerals, which were alienated possibly generations ago to another owner – or retained by the Crown. Or, if you owned the mineral rights, you might sell or lease the rights to oil, but not to other minerals. Or you might find that your property rights are mitigated by access provision to a hydro corridor or to a water body or road – generally called "easements," which have legally become increasingly similar to covenants in Canada, as Marc Denhez, Canada's leading legal heritage authority, explains further in the volume. Among the most important restrictions are those relating to property development – zoning. If you own land that is zoned for single-family dwelling, you cannot build a high-rise. Even more important, there are provincial and federal building codes, which direct how you must build and how you should renovate. Building codes sometimes appear so onerous that heritage property developers give up. Equivalencies, however, can be negotiated that allow both safety and good preservation practice. Larry Pearson explains some of Alberta's standards in his presentation in the volume.

Indeed, real property is often considered as a bundle of "rights," each of which can be sold or restricted in some way. What this bundle might consist of is complex. How might this apply to heritage considerations? The alienation of one or more items in the bundle of property rights could adversely impact the heritage values of a property. For example, the need to run a utility corridor through a property might result in the destruction of a valued historic building, archaeological resource, or landscape. The re-zoning of a property to allow a higher density might well result in the destruction of a precious building. The question that has intrigued individuals and communities who want to preserve their heritage, or at least see elements of it preserved, is whether "heritage" elements are part of the "bundle" of property rights. Can an owner alienate through gift or sale the heritage interests in a property? Can an owner prevent subsequent owners from demolishing the heritage home, from changing the historic interior of a building, or prevent a heritage landscape from being ploughed under? Or can preservation only be accompanied by the intervention of the Crown. As the proceedings of the conference show, the process or the history of the legal considerations are hardly simple.

However, in the instance of heritage property, it is possible for an individual who holds title to the property to commit himself or herself, and more important future owners of the property, to its preservation. The Crown (i.e., the government) as the ultimate owner of all property also has a right to place a restriction on a property, which would prevent the destruction of heritage elements. In many provinces of Canada, that is considered a "taking" of a property right, and while there is no dispute as to the right of the Crown to take these rights, there is as to whether the owner should be compensated. If the owner has lost something through this "taking," he would argue for compensation if there were a financial loss. The owner would argue that one of his or her interests in the property was expropriated. And indeed in most provincial jurisdictions like Alberta, it is probable that the province's Land Compensation Board or its equivalent would determine this a "taking" or "expropriation." The word "probable" is used since the Land Compensation Board, at least in Alberta, has never been involved in a designation issue. Generally the Crown (the Minister of Community Development responsible for the *Alberta Historical Resources Act*) has rarely designated over an owner's objection. Even where it has, it attempted negotiated resolutions without recourse to the Board.

Most important, however, is the question whether an owner can alienate some or all of the heritage rights or interests through gift or sale to another person or charity. Could an owner, for example, agree to never alter the outside of his or her heritage house and bind all future owners? If he did give away this interest in his property, does it have a value? When an individual alienates through sale or other agreement the "rights" to another interest, the parties are construed to have entered into a covenant or agreement. Generally this agreement is to preserve certain pre-determined elements of the property, for example, the front of the building, its roof lines, or one or more of its rooms. The questions that relate to the "covenant" or agreement can become complex. Who can overturn this covenant or agreement? If an agreement were executed between a willing owner and a Trust, could the Crown alter that agreement? If a Trust had given a tax receipt to an owner for the heritage interests, and the Crown changed the conditions of the agreement, would the Trust (whose interests had now been diminished) have the right to be compensated for the loss of its asset – the covenant? If owner of the building and the owner of a covenant were in dispute and the Crown reassigned the covenant to benefit the owner of the building, how would the owner pay for the increase in value? Could a developer who has subsequently acquired the property change the agreement? If the value of the property is diminished by the gifting of the heritage interest, is the giver entitled to a tax break? How would this be calculated? What happens if subsequent owners fail to adhere to the covenant? What if the holder of the covenant fails to enforce it? Can the new owner of the heritage interests then sell them to another party? Do they have any real market values? For how many generations is the covenant binding? It might seem a naive question, but government revenue agencies have a real interest in every one of these questions.

There is a considerable tradition in Great Britain and in the United States, which has found support in the courts to support covenants. But in Canada, despite the successful Ontario experiences, covenants are still in the early decades of experimentation. The environment is one where society expects the state to undertake heritage preservation or to support preservation through generous grants. The covenanting model, one that would allow individuals and communities not only to share with the government the decision-making on the nature of the community, but also to actually take the initiative, is still for the most part foreign. Yet for Canada, whose legal traditions are unique but have roots in the United States and Britain, it does appear to offer a useful tool for individuals and communities to use in shaping the evolution of the places where they live. But, it is not a simple tool.

The preservation/heritage environment in which covenants must operate is uniquely Canadian. To date, in Canada, heritage preservation has been largely driven by provincial statute and regulation.[1] Even Canada's vaunted Heritage Canada Foundation was created in 1973 with a dowry from the federal government, although the foundation has no unique legal powers. Most of the provinces followed suit with their own foundations and legislation, which are listed on the Heritage Canada Foundation website.[2] The federal government had no choice since land, except that owned by the Crown in Right of Canada, is under provincial jurisdiction. Whatever the constitutional reality, in Canadian tradition, which values government intervention in certain spheres, lobby groups and individuals expected their governments regardless of whether federal, provincial, or municipal, to lead. The provinces exercised leadership, and by the early 1970s most provinces had legislation in place. Legislation and regulation quickly became the instrument of choice for Canadians to preserve their heritage. Yet legislation is only as powerful as the will to implement it. Often even if governments have the legislative hammer, they too often did not accompany the "hammer" with incentives – the "carrots." Where governments were slow to act and the carrots either did not exist or were insignificant, their citizens protested and fought demolitions of key heritage buildings. But often they were not successful. And soon developers began to see "heritage activists" as an unrealistic enemy who failed to see the bottom line. Even when they were successful, heritage advocates were painted with the brush of impracticality. They were the enemies of progress. It became increasingly difficult to get heritage activists, elected government members, and developers together to resolve issues. Issues soon became so "professionalized" that much became resolved behind closed doors, the resolution entirely dependent upon power relationships.[3] And important buildings are lost because the fiscal instruments for success are unclear, for example, the Tegler building in Edmonton, or St. Mary's School and the Crown Building in Calgary.

This reliance on the legislation as the solution (the "stick"), particularly when it focused on real property issues, was nowhere well received by the development industry. No matter whether you were a preservation advocate or a developer, all agreed that preservation could be an expensive business. Preservation required money and any active program required more money than any legislature was prepared to vote. In the heyday of budgetary expansions, most provinces established grant programs of one kind or another, and for a few years, at least in the wealthier jurisdiction, they provided solutions. In Alberta, for example, the generous matching grant of $75,000 for Provincial Historic Resources could for many years make the difference between the

St. Mary's High School, Calgary Alberta

success or failure of a project. In rural Alberta, it was an incredibly generous amount; in urban Calgary or Edmonton, it was laughable consideration for major downtown projects. This amount has not changed in over thirty years, making the grant program increasingly unable to meet the challenges of an even larger number of urban projects.

For a number of decades, the apparently generous grant has compromised the need for imaginative community-based or development industry-based solutions. Grants were the solution. There was a belief in the inevitability of conflict between preservationists and developers anyway. There was precious little in the way of policy or legislation that encouraged the development industry or private property interests to come forward and willingly participate in the new initiatives to preserve and promote Canada's past for its future. In the 1980s, however, the Province did attempt to persuade the development industry that heritage preservation was sound economic practice in the long run – that "heritage pays." But it was, and in some areas still is, a wiser investment decision to knock down a heritage building and put up a parking lot, than to rehabilitate it. Canada's taxation system has allowed for terminal loss of property through demolition but provides less advantageous tax treatment for renovation. There are continuing efforts by Parks Canada and by the Heritage Canada Foundation to modify the tax regime, but they have yet to be successful. Even the most recent success by Parks Canada to create a national registry is backed by a modest $10 million in grants rather than by tax relief. And this sum is a one-time allocation, although all are hopeful it will be renewed! The budgets will never be enough. Most of the resistance to a tax-based solution comes from the Canadian Customs and Revenue Agency (CCRA), which believes that special tax considerations are a form of expenditure or "tax expenditures" and that policy is better met by legislatively voted expenditures through grants. While technically they may be correct, there is a psychological and indeed a real difference between applying for a grant and the adjudication bureaucracy that that entails and filing an annual tax form. The taxation process appears less cumbersome and, because it is culturally ingrained, more "acceptable."

The United States and Great Britain, with greater experience in the preservation game, had in fact explored and implemented other approaches, which encouraged private-sector actors to become agents for preservation. In the United States, the federal government brought in a favourable tax and legislative regime, which allowed not-for-profit organizations to enter into agreements with heritage property owners. If listed in the National Register and renovated for commercial purpose, American business might do well. If a private owner wanted to donate property to a heritage agency, he or she could also benefit from very generous tax legislation. While the tax regime in Britain was different, their punitive inheritance taxes drove many of the great estates into the British National Trust. The result was the same – both countries developed a considerable experience with covenants and easements.As mentioned above, the real reason behind Canada's patchwork approach and inability to achieve a national approach to preservation lies directly in its constitutional arrangements and in the construction of its national preservation agencies. Until recently, the National Historic Sites Board of Canada was interested in commemoration, and its bureaucratic arm, the National Historic Sites agency, until money became scarce, saw its role largely as an owner and interpreter. While they had a hand in the creation of Heritage Canada and provided a modest endowment, they did little else to encourage community initiatives. This was felt a provincial responsibility. So the provinces all implemented acts all with limited powers, some with teeth but most without.

More recently, National Historic Sites has realized that this head-in-the-sand approach will not work and that it does have levers at its disposal. While its activities are still very limited, it does enter into cost-sharing agreements with owners of properties. These agreements sometimes carry the weight of easements, but these are few and far between and tend to support sites of clear national significance identified in the National Historic Sites system plan. That plan is so limiting in its scope that these cooperative agreements can never work on a wide scale. National Historic Sites has also been very active in the preservation of Canadian railway stations. The Board, through the National Inventory of Historic Buildings, a unit within National Historic Sites, has the power to mandate the preservation of significant railway stations. When these stations are alienated, many now have covenants placed on them at the time of their disposition. But these remedies are hardly of sufficient breadth to ensure the preservation of anything but the most outstanding examples of Canada's heritage.

One of the early, most powerful provincial acts was that passed in Alberta by the government of Peter Lougheed, anxious as it was in all things to propel Alberta into the modern world. The Act reflected the best of thinking at the time and included wide powers, some of which still are not understood even by the bureaucracy, nor implemented. The Act has served as a model for similar statutes in provinces like Newfoundland and territories like Yukon. Without going into detail, the Act has given the minister responsible for culture the right to issue stop-work orders when a building is threatened, the right to order impact assessments, the right to order remedial action up to and including preservation, and then the right to designate the building. A provincial historic resource, for example, cannot be altered without the permission of the minister, a considerable power, to say the least. But such legislation is only as powerful as the public's willingness to have it implemented. Consequently, in the twenty-six-year history of the Act, there were only two cases where the property was designated over an owner's objection. The Act also made provision for a quasi-independent foundation that might be able to deal in real estate and enter into covenanting arrangements, but the Alberta Historical Resources Foundation never achieved the maturity nor the support it required to become a real agent for heritage preservation; rather, it chose to become a granting agency – a role it performs with great success.

Section 29 of the *Alberta Historical Resources Act* was one of those remarkable sections that, for whatever reason, had escaped the attention of the provincial civil service responsible for the management of the Act, as well as the ingenuity of heritage activists who kept so many of the civil servants on their toes in the 1970s, 1980s, and 1990s. That section of the Act indicated that the minister responsible for the administration of the Act could designate a not-for-profit society as a covenanting agent of the Crown. This would empower a community group, provided they were approved by the minister, to enter into agreements with heritage property owners. It struck a number of people, in a province like Alberta, well known for its resistance to big government, that this might well be a more acceptable instrument than what some perceived as the confiscatory designation powers of the Province. The fine print was important – the minister could overturn a covenant – but nevertheless a covenant willingly entered into and one for which a consideration might be given could not lightly be altered.

This section of the Act prompted a number of Calgarians to come together to form the Calgary Civic Trust in 1998. While the Trust is a not-for-profit society with a federal charitable number and is not a legislated trust, nevertheless, through it bylaws it can hold an interest in heritage property through covenants or easements. Its focus was to be Calgary. The Trust was also interested in more general issues facing the urban environment, particularly issues surrounding liveable spaces and good design. The Trust continues to cooperate with the many organizations already working in some aspect of city building and contributes skills and ideas to the betterment of Calgary. Precisely the objectives of the Trust are to:

- Promote high standards of planning and architecture
- Provide education in Calgary's geography, history, architecture, planning and natural history;
- Secure the preservation, protection, development and improvement of features of historic and public interest;
- Promote civic pride.

In pursuit of these goals, the members of the Trust began to investigate covenants and easements with some vigour. On consulting lawyers and professionals in the heritage preservation field in Calgary, the Trust found there was little knowledge of the covenanting mechanism. At the same time and after considerable negotiation with the Province, the Trust also determined that it could indeed be designated a covenanting agent of the Crown under Section 29 of the Act. A draft agreement with the Crown and a draft covenant were prepared, and the two have been approved. A draft covenant is a part of this book. The Trust had determined that cooperation and surefootedness in purpose rather than confrontation with property owners and developers would the best alternative. The Trust also knew that a favourable tax regime would also be essential.

In working through the complex process of developing the two documents, appended to the proceedings of this book, it became apparent to the Calgary Civic Trust that it would require the very best minds to move the project forward and to gain the support of the development community, the provincial government, the civic government, and the heritage community at large. We would have to act deliberately, carefully, and wisely. The Trust determined that a national conference on covenanting, which had the world's leading experts in attendance (in conjunction with the annual Heritage Canada Foundation meetings in Calgary) would form the opportunity. The precise goals of the workshop were to:

- educate Albertans on the principles of using covenants for heritage preservation and establish covenanting in Alberta;
- enable the public to participate in preserving their own heritage property;
- develop covenanting guidelines and legal documentation;
- and finally to undertake this publication.

The conference attempted to bring the best international experience to the meetings and did so through senior members from the American and British National trusts: Paul Edmondson and Tim Butler. It also brought in the Ontario experience through Jeremy Collins. Their collective approach was critical since the limitations to municipal powers and the reluctance of the Province to designate over owner objections meant that covenanting could be one of the few successful instruments available for heritage protection to a local body. While some would argue that designation is the same as a covenant, it is not. Designation can be done involuntarily, and even if the owner agrees with the action, there would never be a tax receipt. No matter how the designation is construed, it is a *taking*. On the other hand, the willing gift of an interest in a heritage property to the Crown or an agent of the Crown, like the Calgary Civic Trust, should be eligible for a tax receipt. It would be a *gift*, not a *taking*.

Most critical of all was the expertise brought to the meeting on tax issues. To be truly successful, covenanting must be married with tax considerations; that is, the gift of a heritage interest in a property to a not-for-profit must have value. Or why would it be relinquished? The most devastating news during the conference was that brought by Marc Denhez, Canada's foremost heritage taxes expert. In a conversation with Canadian Customs and Revenue Agency staff days before the conference, he had learned that they had decided that heritage covenants, as distinct from easements or covenants on land of natural or environmental value, had only a nominal market value: one dollar. They took this position because they believed that no active market in heritage covenants existed. Their argument is well explained by Marc Denhez further in the book. Basically, CCRA argues that the value of the heritage interest in a property is not to be determined by the "before-and-after" method; rather, it is the value a willing buyer would pay for the covenant. Normally the value of the tax receipt would be calculated by having an appraisal of the property done *before* the covenant was placed on the title and an appraisal *after* with the covenant. The difference in value would be the value of the tax receipt. However, CCRA argues that, since there is no real marketplace in heritage covenants, they have no value. This is despite the fact that Land Compensation Boards are legislated to use the before-and-after method. To reverse this decision will require the assistance of the Department of Canadian Heritage, the lobbying efforts of the Heritage Canada Foundation, and the support of the development and heritage community. There is hope. In the instance of municipal taxation of heritage buildings in Calgary, two appeals have determined that heritage designation has diminished the value of a building to the point where its "real property" assessment has been diminished. It will be only a matter of time before the federal government's CCRA will have to moderate its position as well. Without an owner-driven preservation model, the built heritage interests of this country will always be considered marginal. In contrast, protection of the natural environment, although often contested, has received wide public and political attention and support. Our as-built environment is as critical to our national, regional, and urban identities as are Canada's wild lands and wildlife.

The conference proceedings also illustrate the position of the development community. As Bob Holmes (the former planning commissioner for Calgary) and Harold Milavsky and Don Douglas (both property developers with international experience) illustrate in their discussions, they are interested in heritage preservation and would enter into covenants provided that covenants would contribute to the "bottom line." The estimate provided by experts that covenants could depreciate the value of a property by about 17 per cent would be enough to make a difference. (Some evidence from other jurisdictions that the opposite can occur: that recognition of heritage significance through a covenant can enhance a property's value, seems for now a utopian dream.) Tax receipts for the reduced value are critical.

What will be the next steps? The most critical one will be to challenge the Canadian Customs and Revenue Agency decisions by the various measures pointed out by Marc Denhez in his papers. At the same time, the Trust will be looking into entering into easements and covenants wherever it can. It has accepted leasehold responsibility for and more recently ownership of the McDougall Cairn site in North Calgary to ensure its preservation and integration into a new neighbourhood. The Trust could accept properties as a gift, issue a tax receipt for the entire property, and negotiate a covenant on the property prior to sale. However, these are makeshift alternatives to what should become a more direct process.

Workshop Introduction

Dr. Michael McMordie

As president of the Calgary Civic Trust, I am delighted to be able to welcome you all here this morning to this two-day workshop on Stewardship through Covenanting. I should acknowledge the generous support that we have received for this from the Alberta Law Foundation and the Alberta Historical Resources Foundation. I would also like to express our gratitude for the support from the University of Calgary. A number of us wear two hats, so we tend to come down heavily on the university office staff and telephones and faxes and so on that we can take advantage of. I am very grateful for that. I must acknowledge with gratitude the co-operation and assistance of Heritage Canada, whose annual conference takes place immediately following this workshop. That co-operation made it possible to bring here a number of presenters who are vital to the discussion of the subject of covenanting that we are going to carry through over these next two days.

I should also acknowledge and thank Sally Jennings for the enormously competent and thorough job she has done in getting this together.

I might introduce then, before we begin the session, some people from the Calgary Civic Trust. In front of me are Lucile Edwards, the Vice-president, Frits Pannekoek, Secretary of the Civic Trust, and our patron, Art Smith, who has been a long and active supporter of important public activities, both as an elected representative at various levels of government and as a volunteer active in supporting many community activities. As I have gathered from his presence at the university on a number of occasions over the last few days, he is also deeply involved with the university. Art, of course, has been an important source of encouragement and strength for the Calgary Civic Trust.

The intention of this workshop, as its name implies, is an opportunity to work together to explore the issue. Covenanting is not something that has been, as yet, established in Alberta as a means of heritage preservation. There are a number of aspects of it that seem difficult or perhaps problematic, though the legislative framework does make it clearly possible. We need to do it co-operatively. We need to explore the implications. Covenanting implies a commitment over time into an indefinite future with various responsibilities assumed by various parties. We need to know more about what those responsibilities are and how they can be met. We see covenanting as another string to the bow. It is another way of approaching the preservation, the continued existence, and proper maintenance of valued environmental and particularly historical resources. In the City of Calgary, this means buildings. It is a way that does not necessarily require government designation, and it avoids the confrontations that too often occur when a valued building is about to be sold or demolished and at the last minute a band of supporters of historic preservation gather together to try to do what they can to persuade the owners to change their minds. It is often too late at that point. Governments are reluctant to intervene for all the obvious reasons. Covenanting offers, we think, a co-operative way of moving ahead, hand-in-hand, with the owners of historic resources in a way that benefits the community. We hope it offers significant benefits to the owners as well.

SECTION 1

The International Context

This first section takes examples from the United States and Great Britain. Each has established National Trust organizations and use covenanting as a tool for the preservation and conservation of heritage properties. In Chapter One, Tim Butler describes the experience of the National Trust in Britain in the area of covenanting. This presentation defines conservation covenants and the framework within which they operate and includes a discussion of heritage protection in the United Kingdom. The main discussion is about the practical issues that the National Trust has encountered in its management of conservation covenants. Not all this experience is transferable to the Alberta context, although significant elements will raise some issues. In many ways, the North American model of covenanting is more progressive and flexible than the approach that has been taken in much of the United Kingdom.

In Chapter Two, Paul Edmondson of the United States describes its more flexible approach. He discusses the legal concepts and the development of covenants as a protection tool. He then focuses on the experiences of the United States National Trust as an organization that not only holds covenants, but assists other organizations to hold covenants as well. Mr. Edmondson also makes clear the distinction between this National Trust and that of the United Kingdom.

The British Experience
Tim Butler, Solicitor to the National Trust, United Kingdom

Conservation Covenants

Heritage bodies can tackle the task of preservation and conservation of heritage properties in a number of ways, ranging from simple influence to outright ownership.

Ownership usually provides the most comprehensive defence against the threat of unsympathetic development. But often ownership is not a realistic option: the body may not have the resources to acquire or manage the property, or the present owner, while sympathetic to the idea of conservation, may not wish to move out of the property. How can the heritage body implement a long-term solution in those cases?

Fortunately there is middle ground between outright ownership and simple influence. Without acquiring the property itself, the heritage body can acquire rights over it: rights which restrict the ability of the owner to change the nature of the property, or which positively require the owner to maintain the property in a particular way. These rights are known as conservation easements or conservation covenants – the difference in terminology depends more on legal jurisdiction than on the substance of the rights.

One notable feature of conservation covenants is that they come with their own legal framework. In Alberta, the legal system is founded on the old English common law system, which has quite restrictive rights that apply to properties. Common law has no difficulty with the idea that an owner of a property would make an agreement with somebody else that they (the owner) can or cannot do something on their property. What the common law does have a great deal of difficulty with is the idea that an agreement can bind subsequent owners of the property. Basically, the law only allows enforcement of those agreements against subsequent owners in situations where the person with the benefit of the covenant owns land nearby, and where the subject matter of the covenant is restrictive. In other words, a restrictive covenant means the owner *cannot* do something, while a positive covenant means they *must* do something.

This is a problem for heritage organizations because they frequently do not own any of the land nearby that is capable of benefiting. They will also want to impose positive obligations to maintain the property as well as restrictive obligations not to change it.

The first solution to such covenanting quandaries is specific legislation. The law in Alberta permits specified bodies to enforce covenants even where the following two tests are not met. (The tests are where the person with the benefit of the covenant owns land nearby and the subject matter of the covenant is restrictive.)

The second solution is to use leases, which are not subject to the same restrictions as other covenants. Having entered into a lease, any covenant given by the tenant (as long as it is relevant to the land) can be enforced by the landlord. In the United Kingdom, the National Trust acquires a property and then lets it back to the original owner or lets it on to another party. In this way a much broader range of obligations can be incorporated.

The National Trust

The National Trust was formed in 1895. Its purpose is to promote the preservation of places of historic interest or natural beauty. The twin aims of the organization are preservation and access. The Trust, a charity, is independent of government and it is very proud of this fact. It is regulated by its own Acts of Parliament, passed between 1907 and 1971. It is quite difficult to convey the scale of the activities of the National Trust when you are talking in a country as vast as Canada. The organization owns 250,000 hectares of countryside, 1,000 kilometres of coastline, 183 historic houses and about

20,000 farmhouses, farm buildings and smaller residential buildings. The Trust relies heavily on the support of its members, 2.7 million of them, for its funding both in membership income and in donations and bequests. One important feature of the National Trust is its ability to declare land inalienable. When it acquires land of particular merit, the Council of the National Trust can declare that land inalienable, which means it cannot then be sold by the organization nor can it be compulsorily acquired from it without a special procedure, which involves both houses of Parliament.

Heritage Protection in the United Kingdom

The National Trust is the leading non-government body in terms of heritage protection in the UK. Alongside it is English Heritage, a semi-autonomous body responsible to government. English Heritage, as well as owning and managing properties in the way that the National Trust does, also has a regulatory function.

Smaller organizations and agencies – building preservation trusts – are formed either to look after a specific building or to look after buildings in a particular area. An example that demonstrates some of the preservation issues is the Covent Garden Area Trust. This was a joint initiative by the public and private sectors to safeguard the Covent Garden area in London. They came up with a rather innovative plan involving multiple leasehold interests to make sure the character of the area was preserved.

The first limb of the regulatory framework is planning control, which is quite restrictive in the UK. Any significant development such as major works, demolition, or change of use is going to need planning permission. For properties of particular historic significance, which have been designated or "listed," an additional level of control applies. Groups of buildings of particular historical or architectural merit can be designated as conservation areas. In relation to areas of natural beauty, there are various nature conservation designations. In theory, this is a very comprehensive protection scheme for heritage, whether natural or built. In practice, designation tends to be less uniform than we would wish, and enforcement even more inconsistent – local government bodies often lack the resources or the will to designate and enforce.

The National Trust and Conservation Covenants

Dunham Massey, Cheshire, Great Britain

The National Trust comes across conservation covenants in three main forms. The first is ordinary landowner covenants, which are dependent on the two tests mentioned before, where the person with the benefit of the covenant owns land nearby and the subject matter of the covenant is restrictive. Often these have been inherited with other land that the Trust has acquired. One particular instance where that is likely to happen is in relation to the large, landed aristocratic estates. Often for many years before the estate came to the Trust, the family had been helping to finance the estate by selling off the outlying areas, either specifically for agriculture or else for development. In some cases the form of initial development has been specified, so that any further development would need consent from the family. When the Trust acquires the core of the estate later on, it takes the benefit of those covenants. We see a good example of this at Dunham Massey in Cheshire, where the Earls of Stamford had for many years been selling off, on quite canny arrangements, land for development in the growing urban area of Cheshire.

The second category, and perhaps the most relevant for the topic that we are going to be looking at, are *National Trust Act* covenants. By the 1930s it was becoming clear that the restrictions on our ability to enforce, particularly the need to own nearby land, meant that many areas were not receiving the protection they should. So included in the 1937 *National Trust Act* was a provision which specifically said that in relation to particular types of restriction, on property of particular merit, we can enforce those restrictions irrespective of whether we have land nearby.

That only overcomes one of the tests – the requirement to have land nearby. It still applies only to restrictive covenants. We do not have the ability under this section to enforce positive obligations to maintain, and that is a major chink in the armour.[4]

In practice, although these covenants are used to protect a wide range of properties, including in some instances individual features such as staircases or painted walls, most of them relate to countryside properties. The countryside properties themselves of course will contain buildings in the villages and farms. One good example of the operation of these covenants is Hambleden in Berkshire. This is an estate of 1,600 hectares, classic English countryside. It comprises two villages, several farms, and three historic properties. When the covenant was given over the area, it was a quiet rural backwater, but it is now prime commuter land. The combination of the fact that the area is so handy to London and the fact that the National Trust has done such a good job of preserving its appearance means that the properties are highly sought after. They are only affordable to quite wealthy people who, when they move into the properties, instantly find these charming cottages too small and want to extend them by putting up a garage or a swimming pool in the back yard. So there is constant pressure for change to what had been a relatively unspoilt area. The Trust simultaneously regards the work that it has done in Hambleden, and continues to do, as one of its biggest successes in preserving the area and also as one of its biggest headaches.

Hambleden estate, Berkshire, Great Britain

The third category of covenant that the Trust comes across is Landlord and Tenant covenants. The Trust still uses these quite extensively – acquiring a property, then letting or leasing the property back to the owner or new tenants. This enables the trust to realize the capital value, and get rid of the day-to-day management, but still maintain a high degree of control over what happens at the property.

Specific Issues Relating to Covenants

Subsequent owners: "This is my home."

Among the specific issues that arise in the administration of conservation covenants is the issue of subsequent owners. Given that minds met when the covenant was imposed, with a bit of luck, minds will not diverge while the original owner is still there. The difficulty comes when the property has changed hands four or five times and the new owner does not necessarily have the same philosophical outlook as the original owner. So far as they are concerned, this is their home and they regard advice from the National Trust, or restrictions from the National Trust on what can be done there, as being interference.

It is not an issue limited to residential properties. The owners of agricultural and commercial properties have also got to make their assets work for them. That necessarily leads to changes in use, intensification or alterations.

There is no complete answer. The Trust certainly has not got it right yet, but there are three solutions that are being contemplated.

The first solution is personal contact. Every property that the National Trust owns or has an interest in has a property manager whose obligations include making contact with the people in the area around the property and building relationships. This is particularly relevant in the case of conservation covenants.

The second solution is the production of printed material. On those estates where the Trust has a large number of requests for approvals to carry out changes to their properties, the Trust is starting to produce printed material that explains the philosophy behind the covenants, the benefits they bring to the local area, and the sort of works which would or would not be likely to find favour with the National Trust. The third element is public meetings, used with great care. It is often a bit like going into the lion's den when interacting with a group of people who feel that they are being constrained in what they can do with their properties. Actually to get them together and address them is a bit risky, but it does demonstrate the will of the Trust to listen and make sure that proper account is being taken of owners' concerns as well as the Trust's concerns.

Level of control

Those who have visited National Trust properties will know that the organization has traditionally applied a very high standard to its restoration work. Most of its open-to-the-public properties are immaculately presented.

That kind of mentality does not always sit easily with control over somebody else's property. People genuinely cannot understand why it matters what colour their front door is, or what materials they use for their windows. The solution is better communication to make sure that the Trust is helping people to understand the significance of these features, but it is coupled with something on the Trust's part as well, which is deciding for itself what really matters. Is the Trust taking more control than it needs to? Should it be focusing on those areas that are really going to make a large impact?

Relaxation of covenants

The possibility of relaxing a covenant is open to anybody whose property is subject to a covenant. The individuals must come and ask us if the Trust will either release it, or release it in relation to a specific proposed alteration or change of use.

The question arises whether the Trust should ever agree – is it defeating the object of having the covenant put on in the first place? Some covenants specifically provide for the application for consent in relation to certain types of change. Then you do not have the same philosophical difficulty. The only real question in that situation is "What criteria are applied in saying yes or no to specific requests?" The Trust tends to look at why the covenant was originally imposed and assess how much real impact there will be from the change which is now being proposed.

Of course circumstances do change and, where they do, the Trust will take account of that. One illustration of this would be in relation to social housing. In a situation where the Trust has a covenant that prevents any building at all, perhaps in a rural area, it may be prepared to relax the covenant for low-cost social housing designed for local people. The philosophy is that the Trust has to look at the bigger picture in terms of preservation or conservation work in the countryside and a key part of that is to have a vibrant local community.

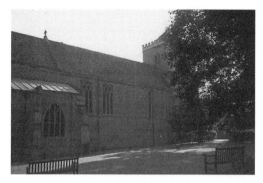

Dorchester Abbey, Oxfordshire, Great Britain

More difficult is the area of absolute prohibitions. Again, the Trust tends not to be too rigid, but not every situation is straightforward. The Trust had one notable case at Dorchester Abbey in Oxfordshire.

In the 1960s a group of concerned citizens who had land around the Abbey clubbed together and each gave the Trust covenants over their land, as did the church, preventing development on the land. That was to stem what was then seen as the tide of commercial development threatening the setting of the Abbey itself.

Recently the church approached the Trust to say that it wanted to make a small extension to the Abbey and, broadly speaking, the extension would have run from the end of the lead roof to the start of the tower, coming out to about as far as the existing structure. The extension was on land over which the church itself had given covenants not to build. The particular difficulty for the Trust in that case was the wording of the covenants themselves:

> No building or other erection shall at any time be erected or allowed to remain on the land or any part thereof it being intended that the said land shall be kept as an open space in perpetuity and not in any circumstances be built upon.

Now as covenants go, they cannot get much more unambiguous than that.

The request, as all contentious issues do, came to the Trust's regional committee responsible for the property. The committee was divided. Some members felt that the covenants were imposed to prevent commercial development and they could not possibly have contemplated the sort of worthy request the Trust was now receiving from the church itself. After all, it was to build restroom facilities, which included toilets for the disabled and other things connected with the fundamental use of the Abbey. The Trust should therefore be prepared to relax it. Others took the more hard line view that the covenant was unambiguous and it would be a betrayal of trust to the wider community to release it in any situation, no matter how painful that might be to the church.

Fortunately, the Trust did come up with a pragmatic solution on that occasion, as it had the benefit of the fact that the covenants were relatively recently imposed. Like-minded groups linked to the church imposed many of the covenants and these groups were still in existence, as indeed were many of the people who had given their covenants privately. The Trust got all those groups who covenanted at the time to agree that they were happy for this release to go ahead. If the covenants had been given a long time ago the position would have been more problematic.

Charging for covenant approvals and relaxations

Linked to the issue of covenant relaxation is the question of charging for relaxation. Is it ever appropriate? What the Trust needs to bear in mind is that not all covenants that have a conservation effect were imposed for the same reason. The Trust has a number of covenants that were imposed specifically for financial reasons because, when the land was disposed of, the development potential of the land could not be fully reflected in the market price at the time. Therefore the Trust included a claw-back provision. I think in those situations the Trust would not have any concerns about enforcing either a specific claw-back provision or effectively doing so by agreeing to the work in return for a payment. More difficult are those covenants that were definitely imposed to protect the appearance of the property. Again, there are two schools of thought which I've rather unkindly characterized as the *purist* view and the *mercenary* view.

The purist view is that if the covenant is there to protect the appearance of the property, then your only legitimate criterion for assessing whether an alteration is appropriate is whether the proposed alteration detracts from the appearance. You should not allow your mind to be clouded by money.

The mercenary view, or perhaps better put, the commercial view, is that if fundamentally the alteration is unobjectionable, and if the effect of the alteration will be to see a substantial increase in the value of the covenanted land, then that means that the National Trust (as the person with the benefit of the covenant) has a valuable right: the right to say yes or no. That right should not be given away when somebody else is going to benefit from it, without the Trust taking a share of that benefit. It would be irresponsible for it to do so.

There is neither a right nor a wrong answer. A lot of these issues are dealt with at the local level, but certainly it is something the Trust should be giving thought to. The Trust should also consider whether it is appropriate for it to give more guidance than it presently does to the regions that have to address these issues.

An interesting illustration of the two types of covenant (those which are imposed on the claw-back basis, and those which are imposed to protect the appearance) is at Dunham Massey, the estate in Cheshire mentioned earlier, where some of the land was sold off to developers on a restricted basis. Either they could carry out only certain development, or they could not develop it at all without the approval of the family. In those situations where developers come and ask the Trust for consent, as a condition of giving that consent, the Trust charges around 1,200 pounds ($2,500) in relation to each new home to be created.

Correspondingly, in relation to *existing* properties where the Trust views the covenants as being there to protect the appearance of the area, the only charge it makes for changes to those properties is a nominal amount to cover fees.

The National Trust's approval procedure

The Trust follows particular processes in dealing with these requests for changes. Minor changes are dealt with by the permanent staff in the region, while more significant changes go to the architectural panel, a non-paid group of architects and other heritage professionals who meet every six weeks. There is a lot of expert input on the panel, but the downside is the potential for delay given the six-weekly meeting cycle. This happens particularly if the decision is that the works are nearly there but they need a bit of adjustment. By the time that has been sorted out, maybe by coming back to another panel meeting, and by the time lawyers have been instructed to make sure it is all properly documented, it can be a long time since the original request. The Trust does not have a solution to that.

One solution that the Trust has broached is the idea of getting involved at the outset so that it can give advice rather than just be there to say yes or no. That works in theory, but it has substantial resource implications. It will not surprise you to know that people whose properties are subject to these covenants are not in a big rush to pay for the kind of additional input from the Trust that this envisages.

Interaction with the planning and listed building system

Linked to this is the interaction with planning and listed building consent, mentioned earlier. There is a degree of confusion here. The planning and listed building process involves delivering all the plans to the local authority, paying the fees, and waiting for a couple of months for the decision. If it is favourable, people in some cases genuinely (and in some cases not so genuinely) believe that if anybody else (other than the planning authority) has the power of veto over their works, they will have been consulted as part of that process. So not infrequently the Trust comes across situations where planning permission is granted and people go ahead with the works, even though they knew their properties were subject to National Trust covenants.

In terms of a solution to that, the Trust is trying to work more closely with the local authorities to make sure each knows when the other has received a request for permission to carry out works. In some areas the Trust is also looking at issuing joint good design guides so that people do not have to reconcile conflicting guidance from the local authority as to what is acceptable with guidance from the Trust. It is not plain sailing. Some local authorities take this attempt on our part to coordinate our requirements as being an attempt to compromise their independence, and are quite hostile to it.

Management implications

Covenanting has significant management implications. Where the Trust takes on owned property, invariably it requires an endowment calculation, and the sums involved can be very large. The Trust does not, and it should, bring the same rigour to covenanted properties. It costs money to look after covenanted properties and it takes time. Considerable time is needed for inspecting the properties. If you have a large portfolio of properties under covenant, even to carry out inspections every six months is going to take up a large part of someone's available working year. Time is needed for liaison with the owners and neighbours, education as to the purpose of the covenants, dealing with the approvals process, and following up on infringements.

Interpretation of covenants

Zennor Head, Cornwall, Great Britain

The issue of interpretation of covenants is a minefield, no less so in relation to conservation covenants. Particular areas of difficulty will centre on subjective qualifications, such as "not unreasonably withholding assent," "minor" works, or non-specific terms like "effect on the neighbourhood," or "causing annoyance."

One solution is to develop a form of standard wording that tackles these issues. It requires a lot of thought and you can iron out some of the difficulties, but it is not a complete answer. Whether the interpretation of a particular document is clear or not is not just a question of the words on the printed page but also of how those words relate to the situation on the ground.

The Trust has an interesting illustration of this at a Trust property in Cornwall called Zennor. Zennor is a countryside property inland from the coast. The covenant there is designed to protect a particular feature, which is a large number of substantial boulders poking out of the ground. The covenant is in broadly standard form and it prevents operations that materially affect the natural appearance of the land. The land is farmland, however, and the covenant was made subject to a fairly standard proviso: that the covenant shall not prevent cultivation of the land in the ordinary course of agriculture. The effect is that one of the farmers who has a farm subject to the covenant is arguing that it is in the normal course of agriculture to take big stones out of your land. This, of course, fundamentally undermines the point of the covenant. At the moment, I think, the Trust has succeeded in persuading him that that is not what it means, but it is a good illustration of how you do need to apply the wording to the particular situation.

Enforcement

The National Trust does not use formal enforcement proceedings very often. Somewhat more frequently, it has to *threaten* proceedings, and that is usually sufficient. The Trust does not willingly embark on formal proceedings partly because it does not like litigation as a method of dispute resolution, and partly because of the uncertainties of interpretation. The knowledge that the sympathies of the court are probably with the property owner rather than with the Trust, and that the court's views on aesthetics may not be as developed (at least that's the way the Trust would like to see it) as ours, make a legal solution fraught with uncertainty.

The Trust also has some confession issues here:

Inconsistent Management:

All too often the offending owner is able to say, "How can you really care about this covenant in the context of my property? Why can't I build this garage? Two years ago my neighbour built an even bigger garage. You never said anything to him."

Delay:

If the Trust does not catch the works before they happen, in practical terms the opportunity for stopping them, or undoing them, is minimal. Where the Trust does catch them in time, it is possible to get an immediate order for cessation. In this case, the Trust must be prepared to underwrite any loss that the owner suffers should it subsequently be found, at the substantive hearing on whether or not there is a breach, that there was not a breach of covenant. The context may in fact be a major new development of a million pounds (a couple of million dollars), which is quite good for focusing the mind in terms of how strong a case the Trust has. We have to consider the consequences of an unsuccessful challenge, since others may take that as an invitation to treat their covenants less seriously as well.

A review of the important points:

- Invest in relationships
- Sell the benefits to incoming owners
- Do not ask for more than you need in terms of control
- Have a clear picture as to what your criteria for relaxation are, and linked to that, whether or not you will be prepared to charge
- Have an efficient approval system
- Develop the appropriate relationship with the authorities
- Make sure you allocate proper resources to the management of the covenant
- Get the wording of the covenant right
- Be realistic about enforcement

I would like to leave you with one thought – that it would be a big mistake to regard covenants as maintenance-free conservation. They may be cheaper than ownership but they are certainly not free. Covenants will not work as a conservation tool unless there is enough time and money available to manage them properly.

How Effective a Tool Are Covenants in the United Kingdom?

The big question is: On the basis of its experience, has the National Trust found covenants to work or not? I don't pretend that question would receive the same answer if it were posed in Alberta.

Conservation covenants are not as widely used in the UK as they are starting to be in some North American and Australian jurisdictions, and I don't really know why that is. I make some suggestions here as to why the UK is different. I think there is a strong proprietorial feeling in relation to homes, and a misguided view that the level of protection through planning controls is sufficient to avoid any threat to a building. Lastly, there are no tax incentives in the UK.

In some non-UK jurisdictions, there are substantial tax benefits allowing you to set off the value of the covenant against income. That is not the case in the UK. There is no discernible tax benefit from covenanting. There is a theoretical benefit when you come to dispose of the property, but I doubt it would influence many people.

Hambleden Valley, Berkshire, Great Britain

The National Trust itself has found that, until recently, covenants were starting to fall out of favour and more attention was being given to leasehold arrangements. That is changing because of the need to look at all possible alternatives to outright ownership.

There are many people within the Trust who regard covenants as unsung heroes. If you look at what has happened at the Hambleden Estate, it would be quite wrong to say that it has prevented change to that valley, but then preventing change is not what conservation is about. The Trust has made the change more gradual and more manageable, and it has been successful in doing that.

A final example shows how well-managed covenants can work. Culham Court is part of the Hambleden Valley property. Culham Court, in the words of the Historic Buildings representative responsible for it, is "a very good Georgian box on the banks of the Thames." It also has impressive gardens and some eighteenth-century parkland around it. It is protected by covenants along with the rest of the Hambleden Valley. Until recently, our relations with the owner were somewhat delicate. But a couple of years ago, the property changed hands. With a new owner who is more sympathetic to our management approach, and perhaps a more enlightened management style on the Trust's part, the property is now being viewed as a fine example of conservation in private hands but with public guidance. Indeed, it is a good example of what can be achieved with a well-managed conservation covenant.

2 The American Experience
Paul Edmondson, Vice President & General Counsel, National Trust for Historic Preservation

This chapter begins with a few words about the legal concepts of covenants and their development as a protection tool in the United States. It focuses on the experience of the National Trust for Historic Preservation of the United States as an organization that holds covenants and assists other organizations in the United States to hold covenants.

First, a preliminary point about terminology. Easements and covenants generally refer to the same type of legal tool in the United States, although the form may not be exactly the same. Here, the term "easement" will be used more than "covenant" throughout the text. The difference is really only significant to scholars of English common law and the vagaries of how that law developed. Because covenanting is done pursuant to a state statutory system in the United States, it is the definition under a particular state's law that is significant. Some states actually use the words "covenant" and "covenanting," but most use the word "easement."

Also as a preliminary matter, we should note the differences between the system in the United States and the system in the United Kingdom. Covenants in the United States are created for protection, as in Britain. However in the United States, covenants also provide economic benefit and tax relief, which is not the case in the United Kingdom. Tax relief is really what drives the system in the United States. It is a significant difference.

Another point of distinction is that, unlike those in the UK, covenants in the United States can impose affirmative obligations. The Trust does not have to resort to leases. Long-term leases are not generally favoured in the United

States. People like to own the land they live on and so covenants in the United States can require an owner to maintain his/her property to certain standards. It makes it a much more effective tool.

A third distinction is that in the United States the National Trust is only one of many preservation organizations. There are approximately 1,500 organizations across the country, with about 1,200 local land trusts that are primarily focused on natural lands and open space. Hundreds of preservation organizations at the local and the state level, and a handful at the national level like us, are able to hold covenants and easements. The focus of our organization's work in the covenanting area is really not on our holding easements, but on assisting other organizations to hold them.

Legal Basis of Covenants

The legal basis for covenanting in the United States is very simple. There is a statutory scheme in every state in one form or another that deals with the covenanting issue. These schemes have common law roots – the issue of the authority of a remote owner to enforce a covenant in perpetuity was an issue under the common law. It has been solved by statute. The practice started in earnest in the 1960s and really boomed in the 1970s. It was enhanced by some early tax court and IRS (Internal Revenue Service) rulings in the 1960s, which were codified temporarily in the 1970s, and provided benefits for the donation of these interests in land. Those tax benefits were permanently codified in the 1980s, and at the same time a uniform model easement law was developed, which provided a kind of cookie-cutter legislation across the country. Virtually every state has some form of this legislation, with very minor differences, primarily as to what types of organizations can hold covenants. In a few states it can be only a governmental agency or a non-profit that is authorized by a governmental agency, but in most jurisdictions any qualified conservation non-profit can hold covenants.

There are conservation easements and covenants covering open space, forestry timbering issues, farmland, etc. Obviously the National Trust focuses on the issue of heritage protection – historic sites – but many easements and covenants these days protect both land and buildings, since most historic sites outside of the urban context have important natural lands associated with them. This is a very significant issue, and one that has led to some controversy and problems of interpretation here and there.

In the United States, a preservation covenant is a non-possessive real property interest that is voluntarily given. In general, this contract has nothing to do with the government. This is a voluntary agreement between two private parties. The covenant is either sold or donated by a property owner to a qualified conservation organization or government entity. The property owner retains ownership, but the covenant controls, in some aspect, the use of land and/or buildings for the purpose of conserving the historic value of the property.

That is the general concept authorized under the state legislation. State laws have a lot of flexibility in the way these things are drafted and the way they actually work. Easements may be for a term of years. You can have an easement or covenant that is five, ten, or fifteen years, or maybe perpetual. It may be what is called "in gross," a remote easement held by an organization that doesn't have property near yours. Or it may be related to a specific property adjacent to the property that is restricted. It may be very restrictive and prohibit any development at all, or may be very loose and have only some slight restrictions. It may be comprehensive and cover an entire property, or very narrow, and cover only a specific structure or part of a property. It may be sold for money, or donated. Private transactions happen all the time for money or personal reasons. Easements may cover structures or land. We have easements to protect archaeological sites – there may not be any structure associated with it – or even historic trails. They can cover exteriors of property – the most common types of easements – or they can cover historic features in interiors. This is universal under state law.

The IRS, in order to assess a tax benefit for an easement, has its own views as to the level of protection required. So there are separate rules that are more restrictive that will enable the owner of a protected property to get a tax benefit in the United States.

- The easement must be perpetual. You cannot donate an easement or a restrictive covenant for ten years and get tax benefit.
- It must be donated to a qualified preservation or conservation organization. The organizations have to meet certain standards as to their conservation purposes; these have to be clear in their charter.

- Private parties cannot defeat the interest. Somebody cannot come along and foreclose on a mortgage and suddenly the easement is released. A private party cannot get rid of an easement; therefore easements meeting IRS standards often have mortgage subordination provisions. In order to get the tax benefits you have to get your mortgage holder to say, "OK, I will subordinate my mortgage to that easement."
- It must be a historically important land area or a certified historic structure. Properties will be qualified if they are listed in the National Register of Historic Places, the federal government's listing system for historic sites and properties. (The national system does not have regulatory impact except for property of the national government itself. There are so many sites and properties across the United States that hundreds of thousands would fall into that category, if not millions.) Properties will also be qualified if they are included in a National Register historic district or a state- or locally-designated historic district that is approved by the federal government, and also certified by the federal government as being of historic significance to the district.
- It must be visible to the public, or accessible in some way. The IRS would not give a deduction or tax benefit if there were no public benefit. That is why many easements that are not generally visible to the public or have interior easement protections have a public visitation provision that says that the property will be held open to the public no less than several days per year. This may be done through a house or garden tour.
- It must be carefully valued. There is a whole system developed for determining the value of easements given up. The most common valuation method is called the "before-and-after" valuation system. Appraisers have developed criteria for looking at a property and saying it is worth this amount in its current state without any restriction. If you consider the fact that it is now restricted and its development potential has been taken away by the covenant, this is what the value is. That difference is tax-deductible.

What are the tax benefits?

The federal income tax deduction may be taken as a charitable gift, just as if you wrote a cheque to a charity. That is a very significant tax deduction. There are certain limits as to how much you can take in a given year. You can only deduct 30 per cent of your adjusted gross income in any given year, but still that is a huge tax benefit, and the deduction can be carried forward. So if you do not use the full amount as a deduction in one year, you can carry it forward five years.

There is also an estate tax reduction. In the U.S., estate taxes are significant, particularly for large tracts of land that have high value – the tax on such an estate can be high. But when the development value is diminished through the donation of a covenant, there is a big reduction in the value of an estate, which means savings that may allow a property to be maintained in a family.

There are some state tax benefits again at the state income tax level (although not a lot of states have estate taxes).

Taking away that slice of value that you are giving to somebody else may also be recognized to lower local real property taxes. Again, that is a significant benefit.

The net effect is a huge incentive for the donation of these easements. These are often combined with other tax benefits. There are tax benefits for the restoration of commercial properties. If a commercial developer takes advantage of the tax benefits, he or she can also roll in an easement donation that increases tax benefits by being able to make a charitable deduction against income from the development (which deduction, if it isn't used up in the first year, may be carried forward for a five-year period). So that is a significant addition to existing tax benefits.

Other economic benefits

Non-profits can actually benefit from donating easements – not through the tax code system – but by giving up easements on their land in exchange for federal or state grants. This is something that we do all the time with historic house museums, when we apply for restoration grants from the federal or state government. One of the conditions of getting those grants is generally to give a covenant on the property. These are usually not perpetual covenants. They are a covenant for a period of years that lets the state or federal authority (that gives you money) have the right to say what kinds of changes you can make to that property.

There are also programs through which state and federal funds may be available for acquisition of easements.

Cost of easements

Despite the simplicity of the concept, easements do not come without cost. Obviously the cost to the landowner is the loss of development rights, but that is a voluntary transaction. The landowner does that either to get a tax benefit or to protect the land. So the landowners are compensated one way or another, either monetarily or because that is what they want to do, so there is some psychological compensation. It is never an unwilling transfer of rights.

The cost to the organization is somewhat different. Money is needed for administration and institutional infrastructure. You need technical preservation knowledge about what kinds of easements to accept. You need institutional policies about types and standards of easements. You need legal assistance – lawyers to draft, to negotiate and to enforce easements. Lawyers, unfortunately, do not come cheap. You need criteria and people to apply the criteria. You need the ability to monitor. We are lucky to get to our properties once every two years. That is the minimum standard we have, and frankly we often do not meet it for some of our remote properties. This subjects us to some danger. The money to cover the costs comes from fees or endowment. Our policy requires an endowment for the accepting of easements in most cases, although there are some exceptions to that where we have partners who bear some of the costs.

You also need enforcement capability, and policies and standards to deal with changed circumstances. I would urge a lot of caution on the issue of approvals for changes to a property after an easement is given, and requests for amendment or relaxation.

Many easement terms are crafted to address a lot of these administrative and process issues. These may include, for example, how an application for approval for change is made and how long our organization or another holding organization has to approve it. Some easements say if you do not approve

Oatlands Plantation, Virginia. Photo: Courtesy NTHP.

it within thirty days, it is approved. This can be very dangerous. It is a larger and more complicated endeavour than a lot of people realize.

The U.S. National Trust Experience

The U.S. National Trust is a very different and smaller organization than the National Trust in the UK. We have far fewer fee or easement holdings. We own only about twenty house museums and a hundred easements around the country. The thrust of our work is to raise public awareness about preservation and especially to expand the capacity of local organizations, our partners, to use easements and covenants.

We use easements in two ways: one is to protect our own properties and the other is to protect other significant properties around the country. I will give a few examples of both categories.

Oatlands, Virginia

Many of our properties have used both fee acquisition and easements to limit development of adjacent land. We started acquiring property around the edges of Oatlands many years ago, and some of our oldest easements are in this area.

Drayton Hall, South Carolina. Photo: Courtesy NTHP.

Drayton Hall, South Carolina

Drayton Hall is a very good "Georgian box" on the banks of the Ashley River in South Carolina. It is a significant historic property and one of the National Trust's most visited sites. The acquisition both of fee interests and of easements around our properties goes on all the time. Just in the last couple of years the Trust has become active in protecting the view across the Ashley River. The land within this viewshed is in private hands and is not zoned; it is on the outskirts of Charleston, South Carolina, in prime "sprawl" country. We convinced the property owner across the way to sell us most of that land for a bargain price. He was willing to let us buy the land provided he could retain some of it for development. We have covenants on the land he retained, with height restrictions on the development. The director of Drayton Hall went over with balloons to see exactly what heights are necessary for protection of the view.

Lyndhurst, outside of New York City

As with Drayton Hall, we are in the process of purchasing an easement next door to the Lyndhurst property. At the same time, we have actually given up an easement on our own property at Lyndhurst, providing an easement to the

Lyndhurst, New York. Photo: Courtesy NTHP.

State of New York as a condition for our receipt of a grant to restore a significant greenhouse structure on the site. It is an easement for a term of years, which we gladly entered into to secure the preservation grant.

Montpelier in Orange, Virginia

Montpelier is James Madison's home: a very significant historic property. We've given up easements for state grant money here. We've also given up easements – this is less typical for us – on the forested land at the back of the mansion to the Nature Conservancy. It is one of the oldest, most significant, old-growth forest stands left in the eastern United States. The Conservancy got the money through a donation from a donor, who got a tax break, and they then used that money to purchase the easement rights from the National Trust. So the tax system has benefited us – we would never have timbered that land anyway.

In all these cases, we were acting as the interested property owner. It is really a more complicated situation in the case of easements that we hold over unrelated properties – in other words, easements that make up our easement collection. But here are some examples of the types of easements that we hold:

Jacksonville, Oregon. Photo: Courtesy NTHP.

Jacksonville, Oregon

Jacksonville is a small Victorian town in southwestern Oregon where we acquired between seven and a dozen easements. This was the start of our easement program in the 1970s. Our program developed really on an ad hoc basis. It became a rather eclectic collection of covenants. A lot of them, however, are on large country properties.

Troth's Fortune, Maryland. Photo: Courtesy NTHP.

Troth's Fortune in Maryland

Many historic houses have significant interiors; the panelling at Troth's is a unique example of late seventeenth-century architecture in the Maryland countryside. Troth's Fortune also has significant open space. Easements at Troth's Fortune were created in the 1970s to protect all three things: the house, the interiors, and the open space around.

Mattapoisett

The National Trust hold an easement on a small saltbox Cape Cod structure in Mattapoisett, Massachusetts, a modest historic property in a fairly isolated location. Why the National Trust has an easement on this property is simply a part of our history. We would never have acquired an easement on this building today – there are lots of state and local organizations that are qualified to hold an easement on this type of property. These small properties can also be a problem, particularly for a remote easement-holding organization. About two years ago a new owner decided that he wanted to put a dormer into the property. As a new owner, he looked at all the next-door properties, which were cute little New England saltboxes but with dormers. He did not understand why this was not allowable – even though we pointed out it was the only house left that was original. He threatened to sue but finally backed down.

By the 1980s, we realized we needed to come up with better criteria for how we accept easements and covenants, and develop criteria and policies. We now accept covenants in two cases. The whole thrust of our policy is to direct easements to state and local entities, which are much better qualified to monitor them. But there are two instances where we do take easements.

The first instance is where people give us properties – the entire property – through our "Gifts of Heritage" program. This is a program through which we accept donations of historic properties, but ones that are generally recognized not to be appropriate for operation as a museum. Our role, in this case, is to use easements or covenants to ensure the permanent protection of the property, but to sell the property to preservation-minded owners so that we can use proceeds for our own programs and sites. This is a very successful program.

Mattapoisett, Massachusetts. Photo: Courtesy NTHP.

The second instance is where we take easements in a "white knight" situation – where the National Trust is the only organization that is in a position to take an easement in order to protect a particular property, or where we have been urged to protect an important property by other preservation organizations.

Lowell's Boat Shop, Massachusetts. Photo: Courtesy NTHP.

Lowell's Boat Shop, for instance, an old boat-building facility in lower Massachusetts, was saved by the National Trust in partnership with the Trust for Public Land. In this case, we were really the only organization that was in a position to step forward and take the easement.

Issues

What kind of restrictions should you put on a property that is literally falling down?

One example of an easement placed on a deteriorated property comes from New Baltimore, on the Hudson River in New York, where we accepted a fee interest in a modest property in very poor condition. What we had to do here is discount the value of the property, considering what an owner or new buyer had to put in to protect it, and so we recognized that we would not get much out of this transaction financially. What we did get, however, was an agreement that required the new owner not only to agree to restore the property pursuant to a restoration agreement, but also to permanently protect and maintain the property through an easement. So there are ways to add fairly strict restrictions to properties, even those in poor condition, that benefit the new owner in some way or another – here through a discounted purchase price rather than a tax deduction.

If this had not been an easement imposed by us on a new owner, but one that an owner gave up to us for tax benefit, the owner also would have been able to obtain financial benefits from a restrictive easement. If this structure were not protected by an easement (it is on the banks of the Hudson and has a beautiful view), it would be worth a lot of money. With restrictions, its potential is much less valued. That lesser value could have been given up through an easement or covenant for tax benefits. So there are ways to restrict a property even when it is in bad condition.

What happens when a property owner does not maintain the property?

This can happen to many different properties. One surprising example is the Kohler Mansion in Riverbend, Wisconsin, which is a fairly large estate owned by a fairly wealthy corporation. The company that owned it really let it go downhill. What do you do? You can threaten to sue. We prefer not to, at least in this context. We do a lot of litigation in other contexts, but we prefer persuasion and embarrassment, and in this case these methods worked. The company is now immaculately restoring the property.

Todd House, West Virginia. Photo: NTHP.

In some cases, however, you have to threaten to sue. The Todd House, a very modest house in West Virginia stands on the site where John Brown, the pre-civil war revolutionary leader, was hanged. The house was in good condition when we were given the property. We sold it and put restrictive covenants on it. The new owners kept it up pretty well for a while, then they had a couple of bad storms and it really went downhill. The owners also made some inappropriate changes, for example, they took out certain protected interior features such as a mantel. We had to threaten to sue and get a local attorney to start sending letters threatening to litigate. That worked. The owner has turned it around and it is now being restored.

Conflicting uses

Edith Wharton's home, The Mount in Lenox, Massachusetts, is a historically significant property. Its tenant is a theatrical company, which uses the setting of the property (which is also protected) as the site for a large outside

The Mount, Lenox, Massachusetts. Photo: Courtesy NTHP.

stage. We had to determine if they had illegally (without authority from us) expanded that stage and, in fact, they had. The current condition of the property is that the stage now covers a much greater area of the property, and the conflicts here have been tremendous. Luckily, these are now ending because the tenant has found another property.

Co-stewardship issues

Easements can be held by two different organizations. The Eastman Hill Stock Farm is a very beautiful property in Maine. When it was given to us, we put restrictive covenants on it and sold it. We transferred some of those covenants to a local land trust in which we shared the interest in the covenant. We are responsible for the structures; they are responsible for the landscape. They monitor the property's beautiful setting and we monitor the house.

There is some conflict in the approaches used by the two organizations on issues such as public visitation and things like signage. For example, we are authorized to put a plaque on the property, but we have had terrible disputes with the property owner about public visitation and the signage. We have very different institutional views on how to proceed. Again, the National

Eastman Hill Stock Farm, Maine. Photo: Courtesy NTHP.

Trust prefers persuasion and we prefer to work with our local property owners. The local land trust, however, has preferred confrontation, which has led to a rather awkward situation. This situation has only improved simply because the ownership has changed.

Case Study: Mar a Lago, Palm Beach

This case has been fun and torture for the lawyers at the National Trust. Mar a Lago is an incredible property in Palm Beach, Florida. It was the estate of Marjorie Merriweather Post, the heir to the Post cereal fortune, and Washington, D.C., socialite. She lived here probably a month out of the year. She would move down for part of the winter, bringing all of her entourage. The house was designed by a prominent architect and then redesigned by a prominent set designer from Hollywood, back in the 1920s. Mrs. Post, when she died, left the property through her estate to the National Park Service. The Park Service, however, declined it, as it was not in very good condition and not the kind of property they really wanted. It was then put on the market by the estate and was bought by Donald Trump.

The property is on open space between the ocean and Lake Worth in the coastal waterway system – hence the name "Mar-a-Lago." The vistas and lawns are incredible, but Mr. Trump wanted to subdivide the property into about ten lots to be called Mar a Largo Estates. The City of Palm Beach prohibited him from doing this. They rezoned it and Mr. Trump sued them. After a while, they settled with Mr. Trump who agreed not to subdivide the property. Under the settlement, the town would permit him to turn it into a club, which was not otherwise permitted, and Mr. Trump would give an easement on the property to the National Trust for Historic Preservation. This is one of those "white knight" situations where there was nobody else who could take it.

This is a somewhat unusual easement in that it is extensively "zoned" to protect different parts of the property – these are set out in a plat included as an appendix to the easement, which defines the critical features of the property. The most critical feature, obviously, is the mansion. The details of the mansion are fully documented through a complete photographic coverage, which is appended to the easement. Each party has a copy of it, so this is an official part of the record.

The landscape is also protected. There are view sheds from the mansion to the lake, and from the mansion to the ocean. There were some drafting ambiguities in the easement that have led to some difficulties in interpretation – for example, the easement protects "the vista from the mansion to Ocean Boulevard," which runs right along the ocean. We soon received a proposal for the Mar a Largo Club cabanas, basically blocking the view of the ocean but not blocking the view of Ocean Boulevard and so arguably permitted by the easement. We were able, however, to invoke other parts of the easement that pushed the cabanas to the side. So the landscape is protected and the list of protections is included in the materials. It includes not just the vistas but also the topographical flow of the land.

Things were quiet for about a year after the easement donation. Then we got our first request for approval and that was for tennis courts. Mr. Trump did not like the original modest tennis courts and wanted to put in a whole new set. There is a tree line on the north of the property that provides a boundary for additional new features, like tennis courts, and we worked to ensure that the new tennis facility stayed within that tree line. We spent probably about four months of negotiation back and forth and finally persuaded to reduce the number of tennis courts from seven to five. We also got him to reduce the size of a service building. The line of trees has been extended so this area is really not very visible – a successful solution.

That was just the first part of Mr. Trump's development ideas for the property. At about the same time, he submitted a development plan for the property. The development plan included a marina. The next thing he proposed was to a ballroom in the area of the lawn on the south side of the mansion. Remember that the easement protects the critical views to the ocean and to the lake. It does not say anything about the views on the side. So he was using this to try to get through a very large structure in what we thought was a very inappropriate place. There are other provisions of the easement, however, that restrict the construction of new structures inside critical features to the extent they negatively affect the mansion itself. We used that as the principal reason for turning him down.

His proposal was basically a ballroom the size of the historic building. We said no and withstood threats by his lawyers to sue. The next thing that happened was that he built a ballroom tent, a temporary structure. The easement, as many easements do, exempts temporary structures. We could not do anything about this. He was not happy with it though, so he kept coming back with ballroom proposals. We tried to convince him to put a ballroom in a corner area of the property which is subject to a very thick tree line. It was too far away from the mansion, however, and he did not like it. We tried virtually squeezing it into every other part of the edge of the property. Finally, a solution was negotiated and we gave conceptual approval for a modified ballroom behind the tree line at mid-property. It had to have heavier landscape protection. This is a good preservation solution.

The moral of the story is – be careful how you draft an easement. The concept of zones is a very useful tool, but you have to keep in mind that, perhaps in five years' time, you may have a Donald Trump as the owner of a property. The next owner may try to exploit every term of that easement language to develop the property.

The other lesson from this is to make sure you get your costs covered. One of the conditions of our taking this property is that it had to be endowed to cover our costs. The endowment given by Donald Trump for this property was about $60,000, which at the time – in 1995 – was the largest endowment we had ever taken. It has turned out, however, to be far too small. This project has generated a huge amount of work for us in terms of staff time, and it could cost more in the future in enforcement.

Amending Easements

I have another case study which involves the whole issue of amendment and relaxation of easements, but I will not go through it because of time. All I want to say is – with emphasis – be extremely careful in amending an easement. We went through a conceptual approval of an amendment several years ago for a property on the Eastern Shore of Maryland known as Myrtle Grove. When we realized that the conservation values that were supposedly being given to us in exchange for letting this property be subdivided (there was a subdivision prohibition in the easement), we backed out of the conceptual approval. The landowner sued us to try to force us to permit relaxation of the easement. We were only able to extricate ourselves from it by getting the Attorney General of the State of Maryland to intervene in the case.

This was, unfortunately, a hard lesson for us to learn, but an important one: When asked to amend an easement, be very careful. Coordinate with interested partners. If you are dealing with a subsequent owner, remember the original donor's intentions, and your obligations to meet them – keep these in mind instead of the complaints of the current owner, who should have recognized the limitations of the easement when the property was purchased.

Conclusion

Like the UK experience, this is an interesting and challenging time for easement-holding organizations: Many easements created in the 1970s and 1980s are now reaching the second and third generations of subsequent owners, many of whom are not as committed to preservation as the original donors, and many of whom want to put their own imprint on their property. More education is necessary – for owners and for the public in general. And preservation organizations have found that they need to be more attentive not just to the built environment but to the whole package: consideration of the environmental setting, particularly for rural properties, protects not only natural and scenic values, but also the overall character of the property.

But the central point here is that preservation easements and covenants have become extremely useful tools for heritage conservation in the United States. They are tools that are being used more and more frequently, particularly in urban areas involving commercial properties. It is important to understand, however, that these agreements are far more complex than many preservation organizations assume: they require the organization to have the technical ability to negotiate and draft easements, the staff to monitor and administer the easement into the future, and the ability to address changes and challenges that may arise with new owners. With attention to these issues, a well-designed easement program can substantially advance the cause of heritage conservation.

SECTION 2

Canadian Precedents

This section highlights covenanting in the Canadian Context. Jeremy Collins looks at the Canadian experience as a whole. In general he feels that the experience of covenanting in Canada lies somewhere between that of the United Kingdom and the United States, leaning more heavily toward the North American context. Mr. Collins looks at the legal background of covenanting in Canada and explains using examples and a case study from the Ontario Heritage Foundation (OHF) with respect to easements.

Next Larry Pearson describes the Alberta context and the newly formed " Protection and Stewardship Section" of the Heritage Resources Management Branch which brings together the department's historical resource management functions into a single dedicated unit. Mr. Pearson looks at cultural resources in Alberta and the preservation tools used to protect them.

Finally, Rob Graham looks at the City of Calgary experience.

3 Ontario Precedents

Jeremy Collins, Coordinator, Acquisitions and Dispositions, Ontario Heritage Foundation

I think you will find that the Canadian experience with conservation easements lies somewhere between that of the UK and the United States, although we are probably a little closer to the United States. In my presentation I will take a look at the legal background for conservation easements in Canada, give you a tour of the easement portfolio of the Ontario Heritage Foundation, and describe a case study based on the Foundation's experience protecting the Aberdeen Pavilion in Ottawa.

Legal Background of Conservation Easements

What is a conservation easement?

A conservation easement is a private agreement between the owner of a heritage property and a heritage body. The heritage body could be a government or non-government organization, or a municipality, depending on the jurisdiction or province. In the agreement, the owner agrees not to undertake any activities that might affect the heritage character and value of the property without the approval of the easement holder. The agreement protects the heritage character of the property. It is registered on title, usually immediately behind the deed, and binds the present and all subsequent owners.

In Ontario, we refer to these agreements as heritage easements or conservation easements. That is probably because Ontario is more influenced by U.S. jurisdictions that use that term exclusively. The term heritage covenants is popular in Alberta, British Columbia, and in Prince Edward Island. The built heritage community tends to refer to these agreements as heritage easements or heritage covenants, and the natural heritage community tends to refer to them as conservation easements or conservation covenants. In Quebec, the term servitudes is used. In my presentation, I will use either easements or heritage easements.

Evolution of heritage easements

The evolution of heritage easements starts with English common law. Heritage easements have elements of both common law easements and covenants. These are historical forms of legal rights with respect to real property, which evolved under English common law, which is judge-made law.

Essentially, a common law easement is "a right of use over the property of another."[5]

A covenant is a promise by one party to another to do something, or not to do something, for the benefit of another party. Covenants affecting real property were only enforceable at common law in limited circumstances. Certain medieval courts of justice enforced covenants provided they were restrictive or negative in nature. Common law easements and restrictive covenants had a unique aspect, which set them apart from ordinary easements when registered in a land registry office or on title to a property. They bound not only the present owner, but also future owners who were not a party to the original agreement.

Historically, there were two main conditions for the enforceability of common law covenants and common law easements. These were that only restrictive obligations could run with the land and bind subsequent owners. Also, the party benefiting from an easement had to own adjacent property that benefited from the easement. To overcome these conditions and facilitate the use of covenants and easements for carrying out a public policy that promoted heritage conservation, there had to be statutory reform. Such reforms have taken place in the last twenty-five years across Canada. I believe that

Ontario was one of the earliest provinces to enact easement legislation, when it passed the Ontario Heritage Act in 1974. Under the Canadian constitution, matters involving private property, including real property, fall under provincial jurisdiction. As a result, it was necessary for statutory reform to come from the provincial legislatures. Each passed laws (or amended existing laws) that would permit the use of easements and covenants for heritage conservation purposes without the common law requirements. The following body of legislation was the result:

British Columbia:	Land Titles Act, Municipal Act
Alberta:	Historical Resources Act
Saskatchewan:	Heritage Property Act
Manitoba:	Heritage Resources Act
Ontario:	Ontario Heritage Act
Quebec:	Civil Code
New Brunswick:	Historic Sites Protection Act
Nova Scotia:	Heritage Property Act
Prince Edward Island:	Museum Act
Newfoundland:	Historic Resources Act
Yukon:	Historic Resources Act

There is no legislation enacted at the moment in the Northwest Territories and Nunavut. It is interesting that most of these Acts deal only with built heritage easements. There are other statutes that enable natural heritage easements. The *Ontario Heritage Act* allows the Ontario Heritage Foundation to take easements for a variety of purposes including the conservation of architectural, natural, scenic, archaeological and recreational resources. It does not limit the Foundation's easement powers to protecting built heritage.

As we review the use of easements across Canada, there are several personal observations I would like to make with regard to the current context in 2000.

The province with the highest activity in conservation easements is Ontario. The Foundation, to the best of my knowledge, operates one of the few provincial easements program in Canada. Several municipalities in Ontario also acquire easements.

Newfoundland has the next highest easement activity. This activity results mainly from the thriving grant program that has been run by the Newfoundland/Labrador Heritage Foundation. In their program, easements primarily protect the public monies that go into restoring the properties and are rarely used for permanent preservation. Owners are at liberty to alter their properties, however, if they do so, the easement requires them to return the provincial restoration grant monies.

In British Columbia, covenants are not widely used. A few municipalities have used them. However, there seem to be two views as to whether an easement or a designation is the preferred preservation tool. Some think that a covenant has no advantage over designation, it is cumbersome, and that landowners in British Columbia are more comfortable with designation because it is a public rather than a private process. Others think that a covenant offers a better form of heritage protection as a mechanism for dealing with landowners and developers in the development context because it can be shaped to reflect the *quid pro quo* dynamic that occurs when property owners agree to municipal covenant conditions in return for planning benefits.

In some parts of Canada, there is a strongly held view that compensation is necessary when property owners' rights are restricted due to heritage designation or easement protection. Sometimes easements may be required as a result of planning conditions or, alternatively, designation protection may be offered to meet planning requirements. In either case, there often has to be some form of compensation. This does not necessarily have to be monetary compensation. For example, in British Columbia municipalities use property tax incentives to promote heritage conservation.

Most provinces have legislation providing for both provincial and municipal designation of heritage properties as an effective form of commemoration and demolition control. However, in Ontario, the current legislation (in 2000) only provides for municipal designation, not provincial designation. In addition, municipal designation does not permanently control demolition but only delays it for 270 days. Presently, twelve municipalities in Ontario have their own legislative authority through private member's bills to extend the 270-day delay period within the Act.[6] However, there are certain conditions that apply and they don't necessarily relate to heritage preservation. Nevertheless, the time delay helps a community to rally heritage interests to seek alternative solutions to demolition. In Ontario, where there has been no permanent demolition control through designation, it seems that more reliance has been placed on heritage easements as the best tool to permanently protect sites from demolition.

Ontario Court case involving a conservation easement

In 1998, there was an Ontario High Court case that dealt directly with built heritage conservation easements. The case demonstrates that there is a strong environment in Ontario for both easements and designations to flourish concurrently.

In the context of a demolition application, this case examined the interaction of provincial legislation dealing with heritage conservation (the *Ontario Heritage Act*) with the provincial legislation dealing with demolition (the *Ontario Building Code Act*). It also compared two statutory preservation tools (heritage easements and heritage designations). Frequently, in Ontario, there can be more than one heritage protection on a particular site, although sometimes this results in confusion. This is exactly what happened in the case of the Roman Catholic Episcopal Corporation for the Diocese of Peterborough v. Corporation of the Town of Cobourg.

Following some delays resulting from the specific municipal heritage designation of its Rectory, St. Michael's Church in Cobourg applied to the Town to demolish the Rectory. The Town's Chief Building Official received the application and approved it because he stated he was not aware that six years previously the church had entered into a conservation easement with the municipality. When he did become aware of it, three months after having issued the demolition permit, the original permit was immediately revoked. The building had not been demolished at that point, although its interior had been taken out. The Diocese applied to the court under the *Ontario Building Code Act* for re-issuance of the permit and argued, amongst other things, that the heritage easement did not apply. It did not specifically cover the Rectory.

In Ontario, an owner is entitled to a demolition permit that complies with Building Code requirements unless the proposed demolition would contravene the *Building Code Act*, the Building Code or "any other applicable law." Since the heritage easement was operative, the court examined the question of whether it could be considered "other applicable law" for the purposes of the *Building Code Act*.

The Court decided in favour of the Town. It found that, while the Rectory was not specifically covered by the easement (the Church was the subject of the easement), the Rectory fell within the lands affected by the easement and these lands were considered to be integral to the character of the Church. It differentiated statutory conservation easements from common law easements noting that the former are public in nature because they play a central role in a statutory scheme for the public purpose of protecting heritage buildings. It recognized that contractual easement agreements and municipal heritage designation by-laws are able to operate as concurrent and compatible forms of heritage protection. It affirmed, however, the provisions of the *Ontario Heritage Act* (section 37 (5)) that if there is conflict between a heritage easement and a municipal designation, the heritage easement will prevail. In light of these findings, the Court held that the heritage easement was "applicable law" for the purposes of the *Building Code Act* and, as such, it was an allowable exception for the municipality to reject a building demolition application.

The case is noteworthy because it stands as what appears to be the only judicial perspective to date on the concurrent effect and operation of built heritage easements and municipal designations on property rights. In affirming the governing nature of a heritage easement over heritage designation status on a particular property, this decision also suggested a possible judicial preference for quasi-private, voluntary restraints on property rights such as heritage easements over public controls on property rights such as municipal heritage designation.

Ontario Heritage Foundation

The Ontario Heritage Foundation is the Ontario government's lead heritage agency and has been in existence since 1967. It was established and mandated by Part II of the *Ontario Heritage Act*. Its mission is to identify, preserve, protect and promote Ontario's cultural and natural heritage.

One of the Foundation's roles is to be a trustee and a protector of built heritage and natural heritage sites. Presently (2000), it owns 117 properties across the province. Of those, 22 are built heritage properties and 95 are natural heritage properties (59 of which are Bruce Trail properties). The

Bruce Trail is a 300-mile trail that runs along the Niagara Escarpment from the Niagara Gorge up to the tip of Georgian Bay in the Bruce Peninsula. The Bruce Trail Association (BTA) was formed in the 1960s with the purpose of securing a hiking trail along the Escarpment for Ontarians to enjoy and appreciate this unique geological formation that stretches across Ontario.

The Foundation owns a variety of heritage sites, including the Niagara Apothecary in Niagara-on-the-Lake. This national historic site has a visitation of about a hundred thousand people a year and is operated for the Foundation by the Ontario College of Pharmacists.

Another of our properties is a natural heritage site called Fleetwood Creek, an area of diverse natural heritage habitat. Water habitats are the core of the Foundation's natural heritage properties portfolio.

The Foundation is also a custodian of cultural, archaeological and archival collections that relate to the sites it owns.

Like the mandate of the U.S. National Trust for Historic Preservation, the Foundation's mandate is very broad. Not only do we act as a trustee, we also act as an educator and promoter of heritage conservation. We accomplish this through a number of activities including the provincial plaque and local marker programs, promoting heritage conservation during Ontario Heritage Week in February, and the Heritage Community Recognition Program and the Young Heritage Leadership Program, which recognize outstanding volunteer efforts in heritage throughout Ontario.

As an entrepreneurial agency, the Foundation raises over 50 per cent of its operating budget from heritage-related businesses and fundraising. Our businesses include conference and reception centres, as well as the Elgin and Winter Garden Theatres – a national historic site and the last operating historic double-decker theatre complex in existence. The centre is rented to theatre and operatic productions. There are also a number of public events, which occur there year round.

Finally, we come to the Foundation's easement program. In 1976, we began our program with our first easement on an historic building. By 2000, we had acquired more than 180 easements to protect a wide variety of heritage resources including architectural, natural, and archaeological heritage as well as recreational settings. With respect to the latter, the Foundation uses its statutory powers to acquire recreational or trail easements to assist the Bruce Trail Association in its trail securement work. Such easement activities are an extension of the Foundation's long-standing partnership with the BTA to secure a permanent hiking trail along the Niagara Escarpment. Under this partnership, the Foundation acquires trail easements over private properties along the Escarpment on behalf of the Association and then it assigns them to the Association.

One component of any easement program is to commemorate those sites being conserved and to ensure that the property owners have a means of signifying that their property is protected by a heritage easement. In 1996, the Foundation began placing small plaques at all our easement sites to address these aspects of its easements program.

How the Foundation acquires built heritage easements

The Foundation acquires easements through a number of scenarios. These include voluntary donations from private landowners who wish to see their heritage property protected in perpetuity, heritage restoration assistance programs that require easement grants to protect public funds, partnerships with other provincial bodies to protect heritage resources, and partnerships with the federal government.

Foundation partnerships with other provincial bodies or with the federal government often involve the easement protection of recognized provincially or federally owned heritage properties which are being contemplated for sale or transfer to the private sector or to a municipality. As an example, the Foundation works with federal bodies to ensure that a heritage property protected under federal heritage legislation continues to be protected under provincial legislation through heritage easements when the property is disposed of and is no longer subject to federal heritage protections. This is the case with heritage railway stations that are designated under the federal *Heritage Railway Stations Protection Act*. These stations are protected under this Act when owned by a federally regulated body such as a railway company. However, when they are sold or transferred by the railway companies, they are no longer protected by the Act but by provincial laws such as the *Ontario Heritage Act* that govern private property. The Foundation and the Historic Sites and Monuments Board of Canada have established a partnership to protect the character of heritage railway stations under disposal by the railway companies. The federal government often places a condition on its disposal authorization that the purchaser or transferee of the station be required to grant a heritage easement to the Ontario Heritage Foundation to protect in perpetuity those features of the station that comprise its heritage character.

The Foundation's easement portfolio

The Foundation's portfolio of built easement properties is a highly representative collection and contains many types of buildings. The largest group is residential properties (at 28% of the easements portfolio), followed by: institutional buildings such as post offices, custom houses, registry offices at 16%; religious buildings of all denominations at 14%; courthouses and town halls at 14%; commercial buildings at 10%; bridges at 3%; natural heritage sites at 7%; railway stations at 3%; industrial buildings such as pump houses at 2%; archaeological sites at 2%; and battlefields at 1%. Twenty-three of the easement properties are national historic sites.

The Foundation holds many built heritage easements on municipal properties. In these situations, the Foundation's easement can play an important role in ensuring that a capital work program intended to upgrade or modernize a municipal heritage building is planned and carried out with sufficient attention to maintaining its heritage character and fabric.

In some instances, the Foundation has used the heritage easement tool to protect the heritage character of a property that it owns and is transferring to another party. One such property was Benares, the inspiration for Mazo de la Roche's "Whiteoaks of Jalna" series. It is a public, house museum in Mississauga. In 2000, the Foundation transferred the property and its collection to the City of Mississauga. However, it retained a conservation easement on the property to ensure the long-term protection of the site.

The Foundation also holds an easement on Dundurn Castle in Hamilton, the home of Sir Allan McNabb, who was one of the founders of the Grand Trunk Railway that ran through Upper Canada.

The Foundation's religious easement properties encompass a cross-section of small rural churches, larger urban churches, and cathedrals. Sometimes an easement protects heritage fabric in a building that is not visible to the eye. The easement on St. Brigid's Roman Catholic Church in Ottawa protects the interior of the sanctuary as well as the exterior. The sanctuary's interior originally had an ornate paisley paint finish. During the mid-twentieth

Dundurn Castle, Hamilton, Ontario. Copyright: Ontario Heritage Foundation

century, the walls were repainted with a plain coat that covered the earlier ornate finish. An area behind the altar was restored in the late twentieth century to uncover the magnificent original finish. One day, the historic finish of the whole interior may be restored. In the meantime, the easement protects

Beaux-Arts CNR Station, Hamilton, Ontario. Copyright: Ontario Heritage Foundation

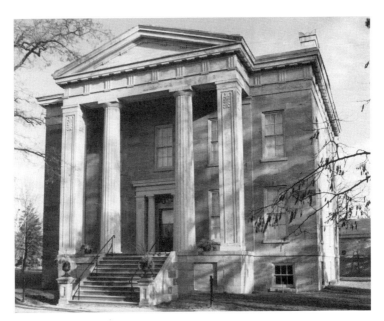

Ruthven Park Estate, Cayuga, Ontario. Copyright: Ontario Heritage Foundation

the heritage character of the exterior and the interior, including the hidden historic finish.

A prominent site amongst the Foundation's heritage railway station easements is the former CN station in Hamilton, a beautiful Beaux-Arts station similar in architectural style and proportion to Toronto's Union Station. The exterior walls possess many bas-relief etchings on a transportation theme. Demonstrating adaptive re-use, the former station is now a conference and banquet facility.

One of the Foundation's most significant heritage easement sites is Ruthven Park Estate, a 1,500-acre estate on the banks of the Grand River, with a large stately home built in 1843 in the Neo-Classical style. The property was designated federally as a national historic site. It had been in the family for five generations since the early nineteenth century. When the second of two sons of the last generation died without offspring in the mid-1990s, the estate passed to an elderly aunt, who was ninety-four at the time. The property was gifted to a local land trust, subject to the land trust granting a heritage easement to the Foundation. The easement covers the whole

Roseneath Carossel, Roseneath, Ontario. Copyright: Ontario Heritage Foundation

1,500-acre estate including a slough forest, farmlands, archaeological resources, the interiors and exterior of the main house and the surrounding landscaped setting. The house and the Estate are open to the public. Most of the furnishings in the main house date to the nineteenth century.

Foundation easements also protect historic battlegrounds. As I mentioned earlier, the Foundation acquires some of its easements when properties are transferred from the public to the private sector. In the case of Battlefield House in Stoney Creek, where one of the main battles of the War of 1812 was fought, the property was transferred to the municipality from a provincial body. The easement protects both the grounds and the interior and exterior of the house, also known as Gage House, which was built in 1796. The easement also protects possible archaeological resources that might be located on the property. The property has undergone extensive re-grading. If there are any archaeological remains from the battle on the site, they are at least twelve to fourteen feet under the present surface.

Finally, one of the most unusual easements in the Foundation easement portfolio is one that protects a historic fairground carousel. A non-profit fairground operator had an historic carousel it wished to restore and was seeking public assistance for its restoration project. The funding body wanted to find a way to ensure that the carousel remained intact in its long-standing location on the fairground. A unique arrangement was conceived to preserve the carousel and protect the public restoration monies using a heritage easement. The carousel sat on a field but it was not permanently fixed on the field. In property law, a structure that is permanently attached to land becomes an improvement on that property and, as a result, it can be protected by a conservation easement that affects land and improvements on the land. The Foundation required the fairground operator to make a cement foundation for the carousel and permanently affix it to the pedestal. Then both parties agreed that the cement foundation would be treated as a built improvement on the property and that the attached carousel was a permanent fixture to the built improvement. That is another extension of property law. When moveable property is permanently affixed to a built improvement (such as lights and built-in furnishings), they are legally considered part of the building and can be covered by a conservation easement. Through this creative arrangement, the Foundation was able to protect the fairground carousel.

Easement Agreements

Contents of an easement agreement

Looking at the Foundation's built heritage easement form[7] you will note the recitals in the initial four or five paragraphs. These tell the story of how the easement interest is created – what powers allowed it to be created, a brief description of the property and, where appropriate, the structures affected. The recitals also define key terms in the easement as well as its scope, clarifying what parts of the property will be covered by it – the exterior, interior, surrounding grounds, etc.

One of the early clauses in the easement relates to approval for alterations and demolition. The Foundation's easement is different from some easements in other jurisdictions in that this clause contains a qualified prohibition, which means that alterations are not prohibited outright but may be permissible with the advance written permission of the Foundation. A previous speaker has mentioned that the sort of easement used in England can be very restrictive. It's rare that that kind of built easement would be used in Ontario.

Another important clause in the agreement relates to access for the easement holder and the public. To be able to enforce the easement, the easement holder needs the right to go onto the property both to monitor it and potentially remedy any breaches of the agreement. There also has to be, as in some of the National Trust agreements, an ability to ensure that there is a reasonable degree of public access to the site. On public sites owned by municipalities and non-profit organizations, we require reasonable public access on a regular basis for public appreciation. For private residential sites, our agreements have a clause that allows for public access once or twice per year under the auspices of a bona fide local heritage group. This reflects the balance between respecting a heritage property owner's right to reasonable privacy and maintaining a reasonable level of public access.

The maintenance and insurance clauses place a positive responsibility upon the owner to maintain the property prudently and carry adequate insurance. There is a lot of jurisprudence as to what signifies a prudent owner of a property. However, to the best of my knowledge, the definition has not yet been judicially tested in the context of an easement situation.

Acquisition process

When the Foundation is contemplating an easement acquisition, staff assess the property in the context of the Foundation's acquisition policies and guidelines and compile a report for the consideration of the Board of Directors. The Foundation has established a set of criteria for assessing cultural and natural heritage easement acquisitions. These include such elements as the significance of the heritage resource, its condition and risk of being lost, its possible connection to Foundation properties, easements or plaques. The Foundation also attempts to ensure that sufficient resources are available to administer the easements in the long term. In this respect, it tries to obtain funds to endow an easement or to secure partnerships to assist it in monitoring or otherwise administering an easement. My recommendation for the initiation of any heritage conservation program would be to look for partners to ensure sufficient resources to carry out the easement holder's responsibilities.

Tailoring the easement to the property

Once an easement acquisition is approved, the Foundation tailors the easement to the specific property and carries out the legal activities, including ensuring that there is clear title. Where there is not clear title, the Foundation seeks to ensure that the party with a prior interest subordinates its interest to the Foundation's easement. It has been the Foundation's experience that it is best to make as few changes as possible to the standard template. By maintaining the standard text and minimizing changes that may suit or be subjective to the present owner but may not necessarily suit or relate to the concerns of future owners, the ability of the easement document to be timeless and affect all future owners equally will be strengthened. One example of this approach is the Foundation's offer of pre-approval process for the purchaser of a property who is contemplating a change soon after the purchase is complete. Rather than include consideration of the proposed alterations in the easement under negotiation, the purchaser submits its proposed alterations to the Foundation in advance of the easement being established. The Foundation assesses the proposal and, provided it respects heritage conservation principles, grants the purchaser and future owner written approval separate to the easement document. The purchaser knows that they can go ahead with the change before signing the easement agreement and the standard easement form is not complicated by an alteration approval that is personal to the present owner.

Baseline documentation report

It is important to complete the baseline documentation report prior to acquiring the easement. The baseline report usually includes basic information about the site, a heritage character statement, and a visual description, usually thirty to forty photographs of the main elevations. This report is the basis upon which the property is monitored for compliance with the easement. Easements are only as good as the ability to enforce them. A good baseline documentation, which adequately reflects the condition of the property from the point at which the easement is established, will make it easier to demonstrate in court that a disputed change occurred after the easement became effective. Typically, property owners are required to acknowledge that they have seen and agree with the baseline documentation report when the easement is established.

Publicizing an easement agreement

It is very important to publicize the existence of the easement. All easements are public record in the Land Registry Office, so any party can see what restrictions exist on the property. However, it is important for heritage organizations to be able to use examples of previous easements to promote an easements program and to promote effective tools for heritage conservation.

Monitoring the easement

Periodic monitoring is addressed in our easement document under a clause dealing with inspection of the property. The monitoring and enforcement aspects of the easements include preparing monitoring reports and determining an appropriate frequency for site visits. Some privately owned sites do not undergo the strains of wear and tear that often accompany public access. Sites that see more public traffic or experience more pressure from an environmental standpoint require, by their nature, a higher level of monitoring. It is important, however, that all easement sites have a regular monitoring schedule so that the condition of the site can be documented over time and to ensure that the easement can be effectively enforced in the long term.

Ownership Transition Information Package

The Foundation has had experience where an easement property has undergone ownership changes and the second owner has not understood the easement as clearly as the first. It is important to get the co-operation of each subsequent owner. To this end, the Foundation developed a package to educate new owners and a package called an Ownership Transition Information Package. The latter allows owners who are selling easement properties to help purchasers understand conservation easements, and to demystify some of the restrictions that occur with easements. Some people see easements as overly restrictive. However, their initial concerns are alleviated when they understand that heritage easements do not contain outright prohibitions and that the Foundation's approach as an easement holder is to work with heritage property owners to find solutions if changes are sought while respecting heritage conservation principles. The Foundation works in partnership with heritage property owners to promote sympathetic stewardship practices for conserving heritage. Regular contact with property owners and an emphasis on building relationships is important to any conservation easement program.

Alteration requests and approvals

With over 150 built heritage easements in its portfolio, the Foundation frequently receives alteration requests. These may include window replacement upgrading and repair, masonry repair, or sign installation. The following is an example of the considerations that sometimes arise when dealing with alteration requests. An easement property owner wanted to carry out capital work by installing single-pane, vinyl windows. The easement site consists of a series of connected row housing and the numerous windows form a key heritage characteristic of this property. After some discussion, an agreement was reached that addressed the owner's concerns for the costs involved while maintaining the integrity of the heritage character of the property. The primary façade fronting on the street was treated with double-sash wooden windows consistent with the original window pattern and fabric, and the secondary façade at the rear received single-pane vinyl windows which were period in style but modern in fabric.

Case Study: The Aberdeen Pavilion

The following case study is an example of the approach that is taken to protect the heritage character of a site with a conservation easement. It is a process that builds on a thorough knowledge of the site's heritage character and fabric to create appropriate easement objectives and ultimately to tailor the

Aberdeen Pavilion, Ottawa, Ontario. Copyright: Ontario Heritage Foundation

easement to ensure the protection of those features that collectively comprise its character.

Heritage Background

The Aberdeen Pavilion in Ottawa is known as the Cattle Castle or the Cow Palace. These alternate names for the building provide a clue as to its original use when it was built as the centrepiece for the Central Canada Exhibition in 1898. Originally a venue for showing livestock, the Pavilion is reputed to be the oldest surviving Canadian example of a large-scale exhibition hall and is perhaps the only example surviving from the nineteenth century. It features a roof structure with a forty-one metre clear span steel frame, with pressed metal exterior cladding. It has eccentric ornamentation with a cupola-capped roofline and elaborate pressed metal decoration of whimsical classical and agricultural motifs. So it is truly an architectural heritage gem as well as an engineering feat of its time. The building has been traditionally viewed from public roads that form the boundaries of the Exhibition Park. This building faced demolition a number of times. Finally, through a combined federal, provincial and municipal effort, an opportunity occurred to restore the dilapidated building to its original grandeur. As a condition of public funding, the owner granted the Foundation a conservation easement.

Easement Objectives

In light of the heritage background, three easement objectives were established to:

- Protect the Pavilion from demolition, and its unique heritage character from unsympathetic alteration
- Ensure that future development within the proximity of the Pavilion was compatible with the Pavilion itself (it now sits in a large parking lot with a few nearby structures, one of them being Lansdowne Stadium)
- Protect existing sightlines of the Pavilion from Bank Street as well as from the road adjacent to the Rideau Canal

Elements of the Easement Protection

The easement protects the exterior of the building as well as the interior heritage features. One of the features that the easement covers specifically is the forty-one-metre, clear span, steel frame which creates the unobstructed interior space and which was relatively new construction technology at the turn of the century. During the restoration, it was important to ensure that at least a portion of structural engineering aspect remain visible to the public. If unsympathetic alterations occurred following the restoration that might obscure the steel span internal structure, the interpretation value of this unique example of Canadian architectural and engineering heritage could be lessened. The easement helps to ensure that the results of the restoration are maintained and are not unsympathetically altered.

The Foundation worked with the owner to create zones around the building with appropriate restrictions to ensure that possible future development would be compatible with the character of the Pavilion and safeguard existing sight lines.

Concluding Remarks

Heritage easements are an important tool for the protection and conservation of heritage properties. Their use differs significantly across Canada. They appear to be more frequently applied in jurisdictions such as Ontario where heritage designation does not provide permanent demolition control. They are complex and technical to use but can be tailored to the unique attributes of a heritage site. With the help of heritage easements, the Ontario Heritage Foundation has been able to permanently conserve a broad selection of provincially significant cultural and natural heritage sites across the Province at a fraction of the cost full ownership would have entailed. In the process, heritage easements have facilitated many partnerships between the Foundation and heritage property owners, as well as heritage partners at all levels of government.

The Preservation and Planning Context in Alberta

Larry Pearson, Protection and Stewardship, Heritage Resources Management

My position is manager of the Protection and Stewardship Section of the Heritage Resources Management Branch, a newly formed branch that brings together the Department of Community Development's historical resource management functions into a single dedicated unit.

The Alberta Historical Resources Act governs the management of the Alberta's historical resources. It provides the basis for all of our activity as a provincial government in managing our heritage resources. The Act defines an historic resource as "… any site, structure or object significant for its archaeological, palaeontological, prehistoric, historic, cultural, natural, scientific or aesthetic interest."

That is a broad range of attributes. It means we can give consideration to a lot of things. The first thing we do in managing the historic resource is to make the determination that it is, in fact, an historic resource.

What Does It Take to Become an Historic Resource in Alberta?

Generally speaking, historic resources obtain their significance in four fundamental ways:

Hillcrest Cemetery, Crowsnest Pass, Alberta

1. They may be significant by virtue of their association with an important event or series of events;
2. They may gain significance through association with a significant person or persons;
3. They may also have intrinsic values:
   - they may be important examples of an architectural style;
   - they may demonstrate a particular aspect of craftsmanship or technology.
4. A site may be significant for its information potential – what it might tell us about our past. This is the case particularly with archaeological sites.

Examples of Cultural Resources

These are examples of the breadth of the resource types that the province of Alberta has sought to protect:

Frank Slide in Crowsnest Pass is an example of a site that has gained its significance through its association with an event – in this case the Frank Slide.

It was designated in 1977 as a Provincial Historic Site. The designation also preserves the site for its scientific importance as a natural site of geological interest.

Head-Smashed-In Buffalo Jump, Southern Alberta

The Hillcrest Cemetery (Crowsnest Pass) was designated because of its association with the Hillcrest Mine disaster, an important event. The cemetery holds the mass grave of the miners killed in that accident. It was Canada's single largest mining disaster.

Head-Smashed-In Buffalo Jump (Southern Alberta) is a World Heritage Site. This site is significant for its prehistoric and scientific interest. It is located just outside Fort McLeod. The kill site is where buffalo were driven over the cliff and the processing areas were at the bottom of the hill where the First Nations encampments were set up, and where the meat processing took place.

Father Lacombe Chapel (St. Albert) is emblematic of two centuries of the Catholic Church in Alberta and commemorates the role played by the

Father Lacombe Chapel, St. Albert, Alberta

Hollingsworth Bldg. Photo: D. H. Brown.

Catholic Church and Father Lacombe in the development of Alberta. The site is operated as a Provincial Historic Site (as are Head-Smashed-In Buffalo Jump and the Frank Slide Centre).

Canada Life Building, or Hollingsworth Building (Calgary) was designated primarily for its architectural interest as an example of a "Sullivanesque" style of architecture, developed by the American architect Louis Sullivan. The façades and a small lobby in the building have been restored.

Rutherford House (Edmonton) was the home of Alberta's first Premier and the first president of the University of Alberta. It is also operated as a Provincial Historic Site.

Rutherford House, Edmonton, Alberta

Holgate Mansion, Edmonton, Alberta

The Emily Murphy House (Edmonton) is significant for its association with Emily Murphy, a member of the Famous Five who obtained status as persons for women.

The Holgate Mansion (Edmonton) was designated as a Provincial Historic Site due to its significance as an example of the Tudor revival style of architecture, and for its association with one of the two developers of the Highlands neighbourhood in Edmonton.

Turner Valley Gas Plant (Turner Valley). A number of industrial sites in Alberta have been designated as provincial or registered historic resources to commemorate Alberta's rich history of resource development. The Nordegg Mine Site, and the Medalta Potteries are other examples of designated industrial sites. The Province owns the Turner Valley Gas Plant and is in the process of developing it as a Provincial Historic Site.

Fort George – Buckingham House (Elk Point): This complex represents the first fur trade posts (Northwest Company and Hudson's Bay Company) built on the North Saskatchewan River in Alberta. They were designated both for their association with Alberta's early fur trade history, as well as for the information contained in their rich archaeological deposits.

The Need for Site Integrity

In addition to associative or intrinsic value, sites should also have a level of integrity to be significant. Once a structure or site has been destroyed, I would argue that it does not make any sense to protect the site. We would not protect the site of the "Tegler" Building in Edmonton. We might commemorate the fact that it was present at one time. When the department looks at a structure or site to determine whether it is a cultural resource, it also asks whether there is enough of the resource remaining to transmit sufficient information about its heritage character and value.

Legislative Tools for Heritage Resource Management

The Act allows the Heritage Resource Management Branch to operate a number of programs that help preserve and protect Alberta's historical resources. The branch has a staff of twenty-eight and a budget of $2,152,000.

Inventory and Evaluation Programs

> Programs to provide technical and planning assistance to the owners of resources;
> Regulatory programs to ensure that sites are protected from the negative impacts of development;
> Funding programs through the Alberta Historical Resources Foundation designed to assist site owners with their sensitive maintenance or development.

The scope of these programs illustrates that good cultural resource management is not just the application of regulations. To be effective it also involves providing professional/technical assistance through advisory services programs and financial assistance through grant programs. It is the wielding of both the carrot and the stick in an effective manner. You cannot just say to someone that by virtue of being designated or by virtue of having a covenant placed on your property, your options for developing that property are frozen. You also have to be in a position to work with the owner to foster a sense of stewardship over the property and to provide a range of advice on sensitive strategies where development or change is required.

Tools for Protection

The *Historical Resources Act*, upon which all our activities are based, provides some very specific tools to manage historic resources. It has also enabled the establishment of our support programs. Section 15 of the act is one of two sections that deal with the process of designation. Ontario, as we have seen, protects its resources at a provincial level through the covenanting process primarily because, at the provincial level, there is no legislation that allows Ontario to designate. In Ontario, the municipalities do the designating. In Alberta, our Act has made it easier for the province rather than the municipalities to designate. Provincial designation is relatively easy, especially when you have a willing owner. Of course, whenever you are covenanting a property, there is an assumption that you have a willing owner.

Our primary mechanisms to protect resources have been Section 15 and Section 16 of the Act. We have over 420 sites designated; about 380 of which are structures. Section 16 is used for sites of provincial significance, while Section 15 is for sites of regional significance.

Registered Resource (Section 15)

The difference between the two levels of designation is fundamentally the degree of protection afforded the resource. A registered resource is protected for ninety days. You tell us what you want to change, and ninety days from that date you can do it. Essentially this provides time to either upgrade the level of designation under Section 16 of the Act to that of Provincial Historic Resource, or to meet with an owner and review alternative development strategies that may be more sympathetic to heritage character. In carrying out these negations we are aided by the availability of financial assistance from the Alberta Historical Resources Foundation. For Registered Historic Resources AHRF may provide up to $25,000 over a five-year period.

Provincial Resource (Section 16)

Under Section 16 of the Act, usually reserved for sites of a provincial level of significance, the Minister's powers are a lot stronger. Owners of Provincial Resources must have the express written permission of the Minister before they can undertake any alteration to the resource. This provides a much stronger position when working with owners to develop appropriate forms of intervention. In addition to wielding a larger stick, Provincial Historic Resources are also eligible for higher levels of assistance from the Alberta Historical Resources Foundation ($75,000 in any five-year period).

In essence, our management of designated resources undergoing development involves on-going discussions with an owner in which we pull out our carrots in the form of grants, and our stick in the form of required ministerial approval. The desired outcome is a development package that meets the owner's needs and also meets the resource management requirements in ensuring that the building's heritage character is maintained.

This recognizes the difficulty in "freezing" a heritage resource. You have to recognize that for a building to survive it has to be used. Resources are under stress if they are vacant and compromises have to be made to allow the resources to be a productive part of the community. We have a stronger negotiating position when we are dealing with a resource that is designated a Provincial Historic Resource. These resources also have a greater likelihood for grant assistance from the Alberta Historical Resources Foundation.

Heritage Areas

The Act also allows the minister to establish heritage areas, which are created by the Lieutenant Governor in Council. The Act also enables municipalities to designate resources.

Archaeological Resources

Section 33 of the Act is a very powerful tool to preserve archaeological resources. The section allows the Minister to require Historic Resource Impact

Assessments if he believes that a historic resource is potentially under threat. That tool has also been used less frequently on the built environment when under threat.

Covenants

Section 29 enables covenants to be applied to heritage resources. The Minister, the council of any municipality in the province where the building is located, the Alberta Historical Resources Foundation, or any historical organization that is approved by the minister may hold covenants. One part of Section 29 says that the Minister may discharge any covenant. I think the minister has some fairly strong powers there. So when it comes right down to the protection of historic resources it is ultimately in the hands of the minister, certainly for a designated resource.

Managing a Protected Resource

The management of designated, or otherwise protected resources does not end with the resources protection through designation or placement of a covenant. Post-protection management involves the careful review of proposed interventions to ensure that heritage character is maintained. In Alberta, this process is guided by the department's Guidelines for the Rehabilitation of Designated Historic Resources. Fundamentally, these are the Secretary of the Interior Standards and Guidelines, edited to fit the Alberta context. These are the basic information we use when we are deciding whether to allow a proposed intervention.

Standard 1. Documentation and Analysis: All alterations to historic properties should be based on a sound understanding of the historical and architectural character of the structure or site being altered. If necessary, detailed research and investigation should be done to identify and document all significant character-defining elements.

Standard 2. Phasing: All rehabilitation activities should be phased and managed to protect the historic fabric of the building.

Standard 3. Compatible Uses: When a change in use is necessary to continue the on-going viability of a structure, the uses proposed should be compatible with the existing structure and grounds such that only minimal changes are necessary. In general, the continued use of a property for the purpose it was originally built is most desirable. It is recognized that the original use may not always be appropriate for contemporary needs. We look for compatible uses. The more compatible the alternate use, the easier it will be to accommodate it without making significant alterations to the building. The best use for a building is to continue to use it for the purpose it was designed.

Standard 4. Historic Character: The distinctive qualities and character of a site or building should be preserved. The removal or alteration of historic features or materials should be avoided.

Standard 5. Historic Period: A building or site should be recognized as a product of its own time. Alterations, which are not based on fact or which try to create an earlier or later design idiom, should be avoided. Understand the character defining elements of the resource. Respect the structure's historic period. Often we get owners of houses saying, "I want to make my house look more historic. I want to put extra trim boards on the veranda or on the eaves of the gable roof." We usually say, "No, that wasn't your building. Your building had this level of history. It didn't have that." So this is a caution against making it too historic. Treat it for what it is.

Standard 6. Witness to Change: Changes to a building or site may have occurred over time. These alterations are evidence of the history and development of the property. Because this evolution may have acquired significance in its own right, significant alterations to a property should be recognized and respected. Respect the structure as a witness to change, and retain later elements. Don't always try to go back to the original appearance of the building, or don't always try to restore it. Recognize that later additions and modifications may have gained significance, either through association or because of their intrinsic values.

Standard 7. Style and Craftsmanship: Distinctive stylistic features and examples of craftsmanship that characterize a building should be preserved and treated with respect.

Standard 8. Repair and Replacement: Deteriorated architectural features should be repaired whenever possible, rather than replaced. Where replacement is necessary the new material should match the original as to composition, colour, texture, design, etc. The repair or replacement of architectural features should be based on a sound knowledge of the original characteristics of the feature, based on historical or pictorial evidence and not on conjecture. If you have to replace deteriorated material, replace it in kind.

Standard 9. Alterations and Additions: Alterations and additions to buildings shall be permitted when such alterations and additions do not destroy significant historical material, architectural or cultural material, and when the design of the alterations or additions is compatible with the size, scale, colour, material and character of the property. We ask that the owners do not destroy significant historic material and that any alterations or additions are designed to be compatible with the size, scale, colour, material and character of the structure and site to which they are being added.

Standard 10. Reversibility of Interventions: When the introduction of a new element or material is necessary to stabilize or preserve a structure, such alterations or additions should be removable at a later date should the intervention fail, without damage to the original fabric of the building.

Preservation Strategies

In Alberta we recognize that appropriate interventions in historic properties can take a number of forms. These can be ordered to reflect the degree to which they impact on a property and to which they maintain its authenticity. The table on page 88 can summarize this notion.

This chart shows that there is a range of preservation strategies. I will go through this list and show you some examples of these being used in respect to designated resources in a way that is consistent with our guidelines. I will talk about stabilization and preservation, restoration and sensitive rehabilitation. In deciding on a specific strategy, it is important to recognize that it is better to preserve than repair, to repair than to restore, and to restore than to reconstruct. Any preservation agency will say that the less you do, the better off you are.

Preservation is a process of applying measures to retain the existing form of the building and the integrity of its materials. It can include stabilization, maintenance, and mothballing.

Most Conservative--Most Radical
Based on the amount of change to the existing state of the resource and the amount of conjecture to make the change

EXISTING BUILDING				RUIN	NEW BUILDING
Unaltered	Altered			Re-Assembly	Copy
		RENOVATION			
		Sympathetic		Non-Sympathetic	
		Retain both appearance and function	New function, earance may or may not change	New appearance, function may or may not change	
To retain as is	To return to a specific period in time				To reassemble original components either in situ or on a new site	To reproduce a copy based on an existing building	To recreate a copy of a no longer extant building based on archaeological and historical research
		REHABILITATION	ADAPTIVE RE-USE	RE-MODELLING			
PRESERVATION	RESTORATION				CONSTITUTION	REPLICATION	RECONSTRUCTION

Preservation and Planning Context in Alberta

Maintenance

The Lieutenant Governor of Alberta's Mansion on the Provincial Museum grounds is an example of one of the two best-maintained buildings in the province, together with the Legislative Assembly. Both are looked after very well by our public works department, Alberta Infrastructure. They are good examples of the best way to preserve a building – by continual maintenance.

Stabilization

An example of the stabilization of a structure undertaken by the department was the restoration or stabilization carried out at Leitch Collieries Provincial Historic Site in the Crowsnest Pass. We were faced with a ruin and we still have a ruin, but we arrested any further deterioration of the structure, recapping the walls and re-pointing the masonry. It was a minimal level of intervention.

Restoration

Restoration is the process of returning a structure or site to its appearance at a specific point in time. The particular period or point in time selected is usually determined by its historical associations or for reasons of aesthetic integrity. Restoration involves the removal of any changes that were made subsequent to the period pf significance, and the replacement of any missing earlier features.

Restoration is more than preservation. You are actually restoring or returning a building to a particular appearance at a point of time. We generally do this when the function of the resource is going to be an interpretive site. Stephansson House is an example of a restored building. We returned it to its appearance at a particular point in its history. In general, it is a process of removing later additions and replacing elements that were there, but are presently missing. The value of a restoration is measured by its accuracy and authenticity. Extensive research and highly specialized expertise are required to accomplish this.

There are a number of designated sites, such as the Wayne Hotel, which have been well restored. Echo Dale, just outside Medicine Hat, involved

Wayne Hotel, Wayne, Alberta

rebuilding of missing verandahs. At Dunvegan, on the Peace River, three windows to the right of a door were later additions and so they were removed and the original window was replicated, based on the other windows still in the building. Deteriorated materials were replaced in kind. If you visit you will clearly see the difference between the new and old material because the new material has yet to discolour.

Rehabilitation

Rehabilitation (preferably "Sensitive Rehabilitation") is the process of modifying an historic site, which is still being used for its original function and involves upgrading the building to meet current building codes. Care is taken to ensure that features, which are historically, architecturally or culturally significant, are retained.

Adaptive Re-Use

Adaptive Re-Use is the process of modifying an existing building to enable it to accommodate a new function or use. In "Sensitive Adaptive Re-use" care is taken to ensure that the building's significant features are retained.

Old #1 Fire Hall, Calgary, Alberta. Photo: D. H. Brown.

Sensitive Rehabilitation

The last forms of intervention in historical resources are sensitive rehabilitation, which may involve adaptive re-use. The process of altering a building, either to meet different standards or to enable a different function, can be managed in a way that still maintains the heritage character of the building.

Old #1 Fire Hall, in Calgary, was derelict for years. The City had it on the market and in the end they gave it to Budget Rent-a-Car and said, "We will give you the building for free, but you have to modify it in a sensitive way and maintain it – essentially restore it." It is a Provincially Historic Resource. Budget Rent-a-Car turned the Fire Hall into their regional headquarters. They put their offices on the second floor. The first floor, which had been used for fire trucks, is the rental office and where they now keep the cars. When you rent a car, instead of getting into a fire truck, you get into a car and drive out through the front doors of the Fire Hall. The front doors are not the originals. The doors were changed quite regularly. This is now a fourth set of doors, so we felt we could not tell the developer to restore it to any particular appearance, but do something that is sympathetic and compatible and is consistent with his needs. We recognized that what was being replaced was not part of the historic fabric. They restored the cupola. The dome on top of the hose-drying tower was missing, so Budget

Ralph Connor Church, Canmore, Alberta

Rent-A-Car restored it. They also restored the pressed tin ceiling in the interior space.

Alterations and Additions

Sometimes adaptive use or rehabilitation projects involve the need for additional space.

The Ralph Connor Church in Canmore, is an example of a large addition to a structure. The church is on the left and the addition is on the right and the two buildings are linked. The addition was designed to allow the church structure to retain its presence on the site. The amount of contact between the two was minimal.

Remuddling, or What Not to Do

On the back page of "The Old House Journal" there used to be a "remuddelling section." It once showed a black and white picture of beautiful house with a bay window and a nice veranda. It was covered with aluminium siding then the aluminium siding was painted to look like the house used to be. Visible in the next picture was a bay window. The verandas, including the bird and its shadow, were "trompe l'oeuil" painted on the aluminium siding. This was someone's idea of sensitive rehabilitation. This sort of thing does not appear in Alberta.

Summary

To recap, heritage resources are managed through the *Alberta Historical Resources Act* primarily through the process of designating the resource, either as a registered site or as a provincial site. Once the site is designated, we follow a set of well-understood standards and guidelines that are used widely across the United States. They are used in every state and we apply them here in Canada when we evaluate proposed changes and do our review and compliance work with an owner. We work hand-in-hand with the Alberta Historical Resources Foundation, deftly wielding both carrot and stick at the same time.

5 The Calgary Context
Rob Graham, Heritage Planner, City of Calgary

What we have to do in the main street of Calgary is to actually implement preservation policy. I will speak specifically about the tool kit. I want to give credit to the Calgary Civic Trust and applaud this initiative because, in my view, this is one more piece in the tool kit that will enable us to protect more, and more specific, portions of our heritage.

Inventory of Heritage Sites

The City of Calgary has an inventory of approximately 450 potential heritage sites. The sites have been evaluated according to a council-approved process that goes back about twenty years. More recently, Dr. Harold Kalman of the Commonwealth Historic Resource Management updated that evaluation process after we found a number of flaws in it. So I think it is a much better tool than it was originally. The inventory enables us to make recommendations to Council on Municipal Historic Resource Designation and to the Planning and Building Department (the Planning Department) on how to deal with development permit applications, building permit applications and demolition permit applications.

Evaluation Procedure

The City's evaluation procedure is based on internationally accepted standards and is similar to the methods of the Province and to the Government of Canada, with some minor wrinkles. The three primary criteria are history, architecture and urban context. Sites are categorized as A, B and C. I must say that I am very uncomfortable with categorizing heritage on those standards. I would say categories A and B make it very easy to dismiss category C. This has been a continuing fight of mine over the past ten years. I am much more comfortable with a register, such as the City of Edmonton has, which says, "If it is heritage, it is heritage." If you look at the 450 buildings on our inventory and the approximate 450,000 buildings in Calgary, you are looking at a fractional percentage of the overall. If UNESCO standards state that we should be saving one to two percent of buildings from each representative era, then you can see how marginal it is when you start categorizing down to A, B and C. It is very difficult when you have to argue a case before Council for a category C building that in my view (and often in the board's view) is just as significant as Category A, but that is a constraint of the system within which we work.

Criteria: History and Architecture

Apartment Buildings

The Anderson Apartments is a wonderful apartment building. Calgary does not have very many apartment buildings. The City of Winnipeg has a wonderful collection of buildings dating from the pre World War I era. We have just a handful and some have been lost recently but the Anderson is the finest, looking at style, type and design.

We consider whether it is notable, rare, unique or an early example of a particular style, its artistic merit, composition, craftsmanship or details. In many cases we consider the builder rather than the designer, because much of our architecture is vernacular. It is equally as important as having a designer or proof of an architect for the project.

Anderson Apartments, Calgary, Alberta. Photo: David H. Brown

We also consider construction technology, which does not really come into play in many cases, although we consider sandstone construction to be an important element. So, if a building has a sandstone foundation, it normally rates higher in our evaluation process.

The interior details and any alterations are also important. A site can start with a hundred points but we subtract points if there have been significant alterations that really detract from the character of the building and make it difficult to restore.

Our evaluation process is such that if a site is significant for its history (and if it is going to be a Category A on that basis) that mitigates in its favour rather than having to have a cumulative point system.

The Rouleau House

The Rouleau House is a humble little house, but very important in the context of western Canada and the Francophone presence here. It is one of five buildings remaining in what was the village of Rouleauville, which was the

Rouleau House, Calgary, Alberta.

French town outside the town of Calgary at the turn of the century. The house was built and owned by the brother of Dr. Rouleau, both came from Quebec. There are French Empire influences on this very humble, literally very little house on the prairie. We think it is ultimately quite fragile in terms of preservation. In terms of its history, this is one of the more significant buildings in the City of Calgary.

We ask, as the province and virtually ever other jurisdiction in the world asks, what significant person, organization, institution, activity or event is associated with the building. Is it illustrative of broad patterns of cultural, social, political, military, economic, industrial or developmental history? Lastly, was it built forty-five years before the present date? The City of Vancouver has a progressive policy in this regard. They will look at anything more than twenty-five years old. There are jurisdictions in the country that consider buildings even younger than twenty-five years. Vancouver has a tremendous collection of modern post-WWII buildings. We don't. It was a stretch to lower it to forty-five years in this case.

Cliff Bungalow School, Calgary, Alberta. Glenbow Archives PA-3538-22.

Urban Context

Looking at the urban context, an example is the Cliff Bungalow School in Calgary in the 1920s and is a landmark in the community. Again, we ask, is it a building that is important to the community? Is it a familiar visual landmark? In this case it sits at the foot of a hill, at the foot of an avenue and is visible to anyone crossing two main streets. When they look down they see this building. In the context of landscape and streetscape, does it contribute to the continuity and character of the street or the neighbourhood?

Category A sites

The Stringer House in Mount Royal, with the bell tower, is a Category A. Just below it is Central Memorial Park, an extremely important park in the centre of the city. The park was set aside by William Pearce in the last century to be a central park. I would say in terms of western Canadian history, it is the most important, or best example of an Edwardian public park extant. The City

Model Milk Building, Calgary, Alberta

worked very hard with the Legion to get it restored. Parks heretofore were not included on the inventory, and this is one of the innovations the Heritage Advisory Board has undertaken in the past ten years. Balmoral School, one of Calgary's sandstone schools, is also Category A.

Category B sites

We treat them identically to Category A when we look at any alterations or changes. We use the provincial guidelines. The Council-approved policy, however, says we have to look at them in terms of the activity that the city encourages. If we are before Council on an issue of protection of a site, we treat category A and B buildings the same. The Norman Block built by Sir James Lougheed, named after one of his sons, is a category B building.

Palace Theatre, Calgary, Alberta

Category C sites

These are much more humble buildings, such as main street type buildings with commercial frontage at grade with apartments above. The Model Milk Building, the home of a dairy in Calgary in the mid-20th century is an example. These are problematic sites. These can be very difficult, as the Council or the Planning Department or the developer can say, "But it's just a Category C."

Case Study 1: Palace Theatre, Stephen Avenue, Calgary

I will now look at a couple of case studies. The first is the Palace Theatre, built in the 1920s by the Allen family, who built a chain of theatres across the country. The first one was on the site of the present-day Convention

Centre. It would be a National Historic site today, if it had not been demolished in the 1970s to accommodate the Convention Centre and the Glenbow Museum. However, this was their second theatre – their grand showcase.

Both the interior and the exterior were substantially intact, although the storefronts had been altered in the 1940s and it had a late 1930–40s marquee placed on it. It had been vacant for six or seven years by the time a development proposal came forward to gut the interior and put in a discount chain.

There was huge pressure to have re-use of this site, which had been vacant on the downtown's major pedestrian thoroughfare for years. We were forced to the conclusion that as no one was prepared to re-use the theatre and no one was prepared to designate it, then façade retention was the best compromise. Let me explain. One of the huge albatrosses we carry is that if the municipality designates a property, it has to pay compensation for loss of economic value arising from that designation. We do not have the wonderful statutes that they have in the United States where the designation of Penn State was upheld at the Supreme Court. Here we must pay compensation, whereas if the Province designates, they do not have to pay compensation. In point of fact, the Province rarely, and I do mean rarely, designates over an owner's objections. I do not believe they have not done this in over twenty years at the time this case came before us.

We looked at it and there was no way we could have taken this to Council to ask them to designate this, because we had Famous Players and their holding company against us. Probably the best way we were going to accomplish anything in this situation was to do a full restoration of the exterior. We had money through the Stephen Avenue Heritage Area Society fund, a granting body set up specifically to fund restorations on Calgary's Historic Main Street. We could make it attractive enough and recognize that we were going to lose a portion of the interior.

Fortunately, at this point, we had a very sympathetic Provincial Minister. We also had a very active local preservation society (the Alberta Historical Preservation & Rebuilding Society) that was just beginning to flex its muscles and literally came out and did a full public relations campaign on this. They were able to secure designation of the site. If it had not been for that grass roots movement, the interior of this building would have been lost. At the time, this was a huge fight and the chairman of the Heritage Advisory Board was in a meeting with one of the developers. The developer turned to him and said, "You have just made one of the biggest mistakes of your life. That building is going to be vacant for the next twenty years."

Within eighteen months, "that building" was fully restored and returned to active use as a night-club/performing arts venue and has done very well, ever since. Interestingly enough, the developer who made that comment later accepted an award from the Calgary Heritage Advisory Board for preservation of this site. Things change very quickly and stories change very quickly in this field. One thing I have learned is that the opening negotiating position is nowhere near the truth.

If you are attending the Heritage Canada Conference, go to its opening night at the Palace Theatre and you will see the inside. There are a lot of insertions that I would call, in dental terminology, bridges. They are all removable. But if you look closely at the original plaster work and the paint schemes, they are original and much of the fabric is intact.

Case Study 2: The Norman Block, Stephen Avenue, Calgary

The building was recovered in the 1950s and what was underneath it: the brick and masonry portion. The cornice and pediment were gone but ultimately restored. We said to the chain that wanted the Palace Theatre, "Don't do it to that building. It's insane – it will cost a fortune and you will have this huge public fight on your hands. Come down the street and do it here. There is nothing inside the Norman Block. It was gutted decades ago. We suspect the original facade is underneath the modern front. We will give you grants to make this attractive to you. We will waive whatever we can in terms of development permit procedures to make this possible." They ultimately bought the argument and the exterior of the building was restored.

The one fly in this ointment is that although restoration of the cornice and pediment are immaculate in terms of their truth to the original, the sanded paint job is wrong. When you see it, it will hit you between the eyes. However, we do not get everything perfect and I have been trying to negotiate ever since with the owners to get the property painted so the whole thing matches.

Norman Block, Calgary, Alberta

Overview of the development process

Development permits for anything on the inventory have to come to the heritage planner's desk. From there it goes to the Heritage Advisory Board for their review and comment and they make comments to the Planning Department. They also have the ability to make comments to Council and to the Province in cases of the highest significance.

The Hyatt Hotel and Convention Centre

Many heritage buildings are under economic threat. The new Hyatt Hotel behind the 100 East Block of Stephen Avenue, directly behind the Lineham Block and the Imperial Bank on the corner is a particular case. If you laid that building on its side, it more than equals all the development on that city block. I want to talk about the fights over this particular development because they were long and hard fought. Many of you may look at it and shake your heads and say, "Well, we really lost there." When I show you the various elements of the problem we were faced with, I think you will recognize that we had quite a victory on our hands over what could have been a huge problem.

Hyatt Hotel and Convention Centre, Calgary, Alberta

The city block had to accommodate a fifty-thousand-square-foot convention hall and all ancillary facilities and the class one Hyatt Regency Hotel. That was a given. That was a done deal when it came to the Heritage Advisory Board and to the Planning Department. The Province designated the building on the corner, so there was no problem there. Nothing else on that block was designated. All were equally important and had no protection on them whatsoever. The developer's opening argument was to save the facades and knock away everything behind. Well, that didn't fly very far. These were hard and long fought meetings. There were five-hour meetings, among the most difficult I have had in my life. When you see the size of the development you will recognize the magnitude of the dollars involved, pitted against the issue and idea of heritage.

The Initial Proposal

The Convention Centre itself had to go on the east end of the block. The west end had to be hotel. The opening position was to drive a Plus-15 (an interior walkway – 15 to 20 feet wide – 15 feet above grade) immediately behind the primary facades linking each of these buildings in perpetuity.

This was such a fast track project they had not had the ability to talk through or think through all the implications. They really hadn't had time to do their homework. In this case, if they had put a Plus-15 across these buildings you would have had peoples' feet walking through the middle of the sec-

Neilson Block, Calgary, Alberta

ond storey windows. Then you would have had to step down three feet into the Doll Block. Bear in mind this all had to be handicapped accessible. They were proposing to reconstruct the building that was originally there, which was a little humble two-storey building, which meant functionally you would have had a Plus-15 midway through the roofline. So you had huge changes of grade on this Plus-15 system.

We use the Province's guidelines and they are valuable, but when it comes to big projects like this you really have to bring out international standards, because they deal with issues such as the philosophy of preservation. The Toledo Charter, UNESCO's charters for historic towns, the Venice Charter, the ICOMOS charter, the Appleton charter, even Australia's charter speak to larger problems like this. In many cases in these meetings we found ourselves literally having to buttress our case, by reading to them chapter and verse from the charters. In many cases I felt like Charles Laughton on the Ed Sullivan Show with the Bible before me, explaining, "If you are proposing to do this … this is what the implications are … this is our historic core … this is the best we have … it is the centre and heart of the city. So what you do here has international ramifications."

One of the points made was that you should not be re-introducing buildings that were gone. Let us be faithful here and let us speak to construction at the turn of this century, not construction at the turn of last century. So ultimately these were new constructions. What you think of them I will leave up to you. I think they could be better, but they certainly could be much worse.

There was another hard fight over the Neilson Block, a very interesting building. It was a furniture store. The first three floors were built in the Richardson's Romanesque style. Calgary was booming in 1907–10 as it is now. They had to put an addition on, which was in the Chicago style, very simplified, planar, also using sandstone and concrete. So the building represented two different periods and was a most important element in Calgary's economic history.

The fifty-thousand square feet of the Convention Centre, we were told, was going to take a little bite out of the corner of this block. We fought that one very long and hard and it was clear that we were not going to win. But there were other pieces of this puzzle we were going to win. The Crown Building was demolished, after great public protest. The facade is all that remains of the Neilson. The fight was to get the Plus-15 to be put in at the same grade as the original second floor so that people would not be looking at feet walking through the middle of these windows, and these are big six-foot windows. The additions at the rear of the buildings were removed. This is an interesting case, talking about tool kits. This is where you get into working at the street fighter level. I learned my trade working for Heritage Canada on the Main Street Program, so I was pretty familiar with horse trading and bargaining and trying to figure out what would make something fly. The Province cornered the developer and said, "Fine, you can proceed with this and this designated building, but what we are asking you to do is maintain fifteen feet of the front of the Lineham to retain legibility of its period." This was to be the grand atrium into the Hyatt Regency and there was agreement on this point that all the floors would be demolished.

Then the developer came back and said they did not want to leave it in situ, but wanted to take it all down. So, then the Board's next fight was, "No, you take that down and you no longer have a heritage building. You have a total reconstruction. If we are going to do this, let's do this properly." So one 8:30 a.m. meeting they came in for a two hour meeting with a sub-committee of the Board and they were adamant. I had found out from them earlier what it was going to cost to retain this, which was the real problem. "Well, it is going to cost us $300,000. We cannot afford $300,000 to maintain this facade in situ." It may have been slightly more. So I said, "All right," and I phoned the Province, and talked to the fellow who deals with grants. I said, "Are you prepared to go to the Historical Resources Foundation and say we need a grant here to maintain this facade in situ? This is a significant site on Calgary's main street, etc. etc." They replied they could do that.

So then I went to the Stephen Avenue Heritage Area Society, which provides grants. I told them we were on record for offering $150,000 for this facade to a developer who owned it before the boom happened in the early 90s. "To get it restored, let's carry this forward and say we will use this to retain the facade in situ." That was fine, the Board was happy.

We went into the meeting armed with this knowledge. "All right, it is going to cost you $350,000 to maintain this facade in situ, and you can't afford it. What is it going to cost you to demolish it?"

"Well, it is going to cost us $125,000 and we simply can't afford that additional amount."

I said, "Wait a minute, the Stephen Avenue Society is on record as saying they will put in $150,000, and the Province say they will put in $75,000. There is $225,000. You subtract what it is going to cost you, and it is pretty well a wash." Well, faces just dropped. They could not believe they had been short circuited like this and they stormed out of the meeting. The Director of the Planning Department stood behind us on this and the facade stayed in situ.

That is horse-trading. It has nothing to do with any tools. In that case what worked was a couple of little grants and the fact that we had a publicity-adverse owner, who did not want to get into a big ugly public dog fight over this. They also had a very short time-line and they recognized there were enough hurdles here to go through already. It shows how little real regulatory ability we have to protect buildings. Ultimately, the Plus-15 came across here and plugged in through this wall and then proceeded along through a side wall here and then went to the back of all the buildings, where it should go, and where the original lane functioned. From there it goes into the hotel.

So that was the big win-win, and I think if you walk through the Centre and then look at it from the street you will recognize what a success that was, given overall the challenges we were facing.

Most work on Stephen Avenue is facade restoration. We do not take development permits for it. We do it under building permits. It is faster and cleaner, and it is not a roadblock for the developer.

Quick Summary of the Building Review Process

Demolition

Here we get into the ugly stuff. All demolition permits have to be reviewed by the heritage planner. This is thanks to the Alberta Historical Preservation & Rebuilding Society, a grass roots agency. I spoke to you about it earlier in connection with the Palace Theatre. We did not have that provision before but they pressed that issue on us when a house was accidentally demolished in Mount Royal, two days before it was to be placed on our inventory. There was considerable resistance in the department to this.

I, or my assistant, have to look at demolition permits everyday and we see everything. If it is a garage, we see it. Anything that is demolished in this city we have to see the permit. We do catch things, and in some cases, we can save ourselves a lot of grief, but it is also a very labour intensive process. All sites on the inventory are referred to the Heritage Advisory Board. In cases of highest significance we refer them in turn to the Province. The Law Department has said we have thirty days under the *Uniform Building Standards Act* to take action. They are not discretional. If you apply for a demolition permit, it is not discretionary. If you fulfil the requirements of the Act and you remove servicing, you can demolish.

So we are literally putting a stumbling block in front of the Act. The Law Department is comfortable with thirty days to hold something before it can be demolished. In that time, if we can get it to the Province and get the Province on record as saying, "Yes, they are looking at considering its historic significance," then we can delay it even further, to make sure if there is any window of opportunity to protect. Sometimes in that interval, something comes up through the grass roots, a developer changes their mind, or the economic situation of the developer changes, so that we are able to engineer some form of protection. In some cases, the Board will ask Council formally to ask the Province to designate. We are less successful there, as Council feels it is circumventing the municipal designation requirement of compensation.

The Lougheed Building/Grand Theatre

The "Cause célèbre," which many of you may be familiar with if you read the *National Post*, is the Lougheed Building, built by Sir James Lougheed, which houses the Grand Theatre. It was one of the great theatres of the pre-World War I era in Canada.

This building just looks like a simplified Chicago style with a bit of classical detailing and not much else. It had a rather wonderful cornice at one time that went the length of the building. It was a huge cornice that was ripped off by the owner. It is a Category A site. It was originally a Category C. When I looked at it on the inventory when I first joined the City, I thought there was something wrong here. This is a really significant site. The Grand Theatre is

Lougheed Building, Calgary, Alberta

not even acknowledged. So we had it looked at closely again and re-evaluated to a category A.

At this time, coincidentally, the owner discovered that there was some deterioration within the cornice. We came in one Monday and found they had had it ripped off without a permit. We referred this to the Province who issued a Historic Resource Impact Assessment on this site, which freezes development rights and requires the developer to prove whether or not it is provincially significant. Also missing is a lovely wrought iron and steel glass canopy over the main entrance and another one on the other side of the building.

The Grand Theatre is where Sarah Bernhardt played. Fred and Adele Astaire danced there, Jack Benny played there, Amundsen spoke there, and Paul Robeson sang there. It is an incredibly important site. That said, all that is left of the interior is the second floor balcony, the balcony rail and fragments of the plaster ceiling. The rest of it, with the possible exception of some indoor murals above the second balcony, is gone. It is a 1960s grey Cineplex Odeon box. But it is an extremely important building in terms of the streetscape. It forms part of what some groups in the city call the Northwest Passage. It is the street that crosses Stephen Avenue running north and south and has a number of important early commercial buildings on it.

Ultimately the Historic Resource Impact Assessment was filed. The Minister took a very long time to deal with it – over a year. Finally, that Minister chose not to deal with it, and took another portfolio. The next Minister came on board and looked at it and decided that he was not prepared to designate this and he so informed the city. This meant that a development permit application that had been filed for the site was again active, which was predicated on demolition.

The Heritage Advisory Board then went to Council and requested its purchase or any other option, including municipal historic resource designation. The problem was the compensation clause. The owner was asking for approximately $9 million. We had a separate assessment that came in much lower than that, but what the owner was going to argue was lost opportunity.

Let me explain. If the City and owner cannot come to an agreement on the appropriate compensation, it goes to the Land Expropriation Board. The Land Expropriation Board rarely hears issues of this kind. The City is then compelled to pay all costs associated with that appeal plus the owner's costs and then accept whatever judgement the Land Expropriation Board delivers. Council declined and we again have an active development permit that will allow a demolition permit, and subsequent demolition of the site. This is, of course, unless the Minister changes and unless economic situations change.

The plus side of this is that we believe it is, in fact, not a real development. This was really done by the owner to prove that development could occur on this site. Our best analysis, from the best real estate assessor we could find in Calgary, and one of the best in western Canada, said this is not a real development site in the short term, maybe in the medium term, but not for a long time. So that is the plus, but at this time we still have to process the development permit, which will then predicate that there can be a demolition permit and that we would have to sign off on it.

St. Patrick's Roman Catholic Church, Midnapore, Calgary

If the Province designated this site, it would not prevent further development. If the Minister allowed it, demolition of most of the structure of the building with a high-rise construction behind could be the result.

What would a covenant do for this site? I am asking you and I am looking forward to the discussion tomorrow on this subject. If a covenant were placed on this building preventing construction over six storeys, or perhaps a penthouse, or did not allow construction over seven stories, then if you apply Marc Denhez's before-and-after scenario (if the government of Canada altered its current ruling on donations and tax receipts) there could be some real incentive for the owner, or another owner at another time, to retain this building on site.

St. Patrick's Roman Catholic Church, Midnapore

This is a humble little wood frame building on what was once the edge of the city, on what was the historic McLeod Trail between Fort Benton, Fort McLeod, Fort Calgary and Fort Edmonton. It sits cheek by jowl with a designated Anglican church of the same vintage, same style and same form of construction, with graves on site from early settlers. This is a very important site. The Roman Catholic Church has been trying to sell it for many years. Some years back they attempted to sell it to an auto dealership. The community got word of this and realized the graves would be disinterred. So they went to the Province to get the graves formally protected. All went quiet, and then came back at us about six months ago, when the property was sold to a memorial society who proposes to build a columbarium on the site. Currently there is a request from Council for the Province to designate this site. The Province has indicated it is worthy of designation. That is where we are now. We withheld the demolition permit, but what would a covenant do in this case?

I am willing to argue that if there were a before-and-after scenario in place for covenants, and Revenue Canada were willing to accept a tax receipt from the Calgary Civic Trust (with the Alberta Historical Resources Foundation, or the Calgary Heritage Authority or whoever held the covenant) there would have been enough money to satisfy the owner, the Roman Catholic Church and the Calgary Diocese, who were interested in getting a return on that land. The graves are now protected and will be retained in situ. I do not know what will ultimately happen there, but this is an example of the necessity for another tool in the kit.

Density transfer

One of our tools is density transfer. The Calgary Oddfellows Temple had the ability to sell its air rights in order to protect it during the oil boom of the 1970s. If you look at the buildings around it you see the huge economic threat. Subsequently, a policy was put in place that allows for the transfer of such rights to a willing buyer within the same land-use district. In this case it was PetroCanada. They bought the air rights in the order of a quarter of a million dollars, which then enabled the Calgary Chamber of Commerce to restore the building, use it for their offices and return grade to commercial use. The problem is, in fact, that it is not a huge incentive in a hot economic climate.

Option one in the Core Zone One, the downtown commercial core, is the provision of a maximum density of seven times the floor ratio. An

Oddfellows Temple, Calgary, Alberta

additional three can be added or transferred for heritage preservation purposes. That is not very much when you are looking at a bonusing system, which provides for densities of 15 to 20 FAR in exchange for various amenities. For with densities over 20 FAR, Council can approve on a case-by-case basis. Density transfer is not an attractive tool at this point. The Planning Department is currently looking at retooling that bonusing system, so that it is a better tool to protect heritage.

The Art Gallery of Calgary, formerly the Burns Meat Market – Pat Burns was one of the Big Four who started the Calgary Stampede – and the Calgary Milling Company on Stephen Avenue also sold their rights. They had the ability to sell only two floors because we have sunlight guidelines on Stephen Avenue. You cannot build anything higher than four stories on the south side of Stephen Avenue, because you would shade the street. To get density transfer you have to accept municipal historic resource designation and accept that as compensation. Again the figure was in the order of a quarter of a million dollars, which is not very much money unless you are a not-for-profit organi-

Art Gallery of Calgary, Calgary, Alberta

zation. It is not very attractive at this time. In this economic climate it is an increasingly smaller and smaller carrot.

"What would be the impact of covenants on a site such as the Clarence Block?" It is one of the few buildings left on Stephen Avenue that is not restored at this time. It has a direct control land use designation dating from the oil boom of the 1970s, which would allow construction of a high-rise tower that would extend north of this building. How could a covenant be creatively used on this site to protect the character of the building? There is nothing left of the original fabric in the interior of the building. It was the offices of Sir James Lougheed and Prime Minster Bennett, partners at one time. What is important here? Is it the interior spaces, the interior volume, or a sense of the original structure of that building?

Another case study that I think is interesting is Central High School. A number of inner city schools now sit on huge pieces of property, and they are being de-accessioned. This school is one of them. What will happen to that site? It is in an area that is highly desirable for condominiums and inner city densification. What are the trade-offs here? What are the ultimate uses that could be placed on that site and still retain its heritage character?

Clarence Block, Calgary, Alberta

SECTION 3

Financial and Legal Considerations

Marc Denhez, Canada's pre-eminent authority on tax and heritage assesses the rapidly evolving tax environment. Jason Ness brings the issues of appraisal to Calgary.

Tax Implications: Financial and Legal Considerations

Marc Denhez, Heritage Lawyer

A speaker once said that the definition of an expert is an SOB from the East with slides. I should advise you that I have brought no slides for this particular presentation but, I am a lawyer from Ottawa – nobody is perfect. I have the lugubrious task of defining for you the tax treatment of easements and covenants. It is a dirty job but somebody has got to do it. The question of the tax treatment of these agreements may be absolutely crucial to the ability to negotiate them with private owners.

This is one of several provinces, in fact almost every part of Canada, which has enacted legislation specifically designed to enable the negotiation of these kinds of agreements. You may also be getting the impression that aside from examples such as we have seen and heard from Ontario, these kinds of agreements are like hen's teeth in this country. I believe that the reason for that is extremely simple: there has previously been no Calgary Civic Trust to organize a meeting like this. What you have heard this morning is, in my opinion, the most comprehensive blueprint as to what an easement and covenant strategy can and should look like in any jurisdiction. I think you have probably heard as complete a listing of the concerns that any initiative like this should take into account.

Before getting into the specifics of the tax treatment of easements and covenants, let me just recap some of the things you have heard this morning as to the basic nature of these agreements.

You have heard that these are contractual agreements. They are however distinct from normal contractual agreements, because you will remember that in the case of a normal contractual agreement, it binds the people who sign the agreement. If you sign any kind of normal contract, and the day after you get run over by a bus, your heirs are not normally subject to that agreement. The agreement is binding only on the people who signed it.

This does not work very well in the case of private property, because private property has things happening to it. Private property gets bequeathed to people after people die. Private property, even more often, gets sold to people. So what happens if you want to enter into an agreement that is binding upon people in the future, not just the people who signed the agreement? Well, the answer to that is when you cross from contract law into property law, there are new opportunities that open up.

For hundreds and hundreds of years there have been these specific kinds of agreements that existed in common law, recognizing that there are certain property transactions that can be binding well into the future. One of these agreements was called an easement, and an easement was fairly straightforward. An easement was what you signed when you wanted to allow somebody else to do something on your property: run a hydro utility cable, a pipeline, or various other things where you retained the title to the property but somebody else had the right to come onto your property and do something there.

Then you had covenants. Covenants did not allow anybody else to come onto your property and do anything there. Covenants said, "I, personally, the owner of the property, hereby promise that neither I nor any of my successors, either the buyers of the property or my heirs, will ever do anything here. Totally negative, totally restrictive – I promise not to do anything on my property.

So, by traditional, common law standards, if you signed an agreement that said, "I shall never alter, demolish, or build on my property," you were signing a covenant. If you signed an agreement that said, "If my property is damaged, I will allow somebody else to come onto my property to repair it," you were signing an easement.

What happened if you signed an agreement that said both? "I promise never to damage my property, but if my property does become damaged, I will let somebody else come onto the property, take a look at it and make recommendations as to what I should do, and I will do it, or they will do it for me." Well, what do you have then? You have a mule!

A hundred years ago, lawyers could get really upset about this because it was two thirds of a covenant and one third of an easement, or was it two thirds of an easement and one third of a covenant? Then you were forcing people to pay money in the future, because that made it positive instead of negative, and there was no other property that was going to benefit by this agreement – condition, condition, condition. So the legislature came in and swept it all away.

Now that you have figured out the difference between a covenant and an easement, forget about it, because the statutes of Alberta have said, in the *Historical Resources Act*. As they have said in umpteen other provinces across Canada) whether you call them covenants or easements, they will be binding in the future. This is regardless of whether you have to spend money, regardless of whether there is another property that increases in value courtesy of this agreement, as long as they meet certain statutory conditions. Hence, the result is the covenant provisions of the *Alberta Historical Resources Act*. So this is the animal we are dealing with.

Those of you who live in Alberta know that the process of owning property is a bit complicated. You know that property can be divided not only horizontally (if I go one metre here, this might belong to somebody, if I go another metre here, this might belong to somebody else, and so on) but property and property rights can also be divvied up vertically. If you own the soil, it is not necessarily the same as owning the subsoil. It does not stop there. Property rights in the English common law system are made of a wonderful jumble of various rights, and I am not using the word jumble – it was actually Oliver Cromwell who used the word jumble to describe English common law property. As a matter of fact, he referred to it as an ungodly jumble, but he thought a lot of things were ungodly.

So what happens when you take some of these property rights and you hive them off, or take bits and pieces of them? If they are in one of these traditional, recognizable categories of property rights and you hive them off, you are engaging in a disposition of property which is, in the eyes of the law, every bit as real, every bit as recognizable, as if you had sold off the back forty. This is what you get when you have an easement or covenant.

Let me just recap another couple of things. These agreements, as I said, have been around for centuries. They are, in various jurisdictions, very well drafted. They are tightly drafted. They do exist. In point of fact, in the province of Alberta, easements of various kinds have been around for a very long time, much longer than there has been an environmental movement, just as in other parts of the country.

What do you think happens when somebody acquires a property right that can be registered at the Land Titles Office, binding upon future generations, but is not absolute ownership? It is less than what we lawyers call fee simple. You have less than fee simple property rights. Less than fee simple property rights have been around for a very long time.

As I alluded to earlier, every time the electricity company wants to put an electric cable over your property, or, for that matter, pipelines up, down or sideways, and various other things that cross your property or do things on your property – these are all easements. Everybody in this province, and for that matter across the country, is very familiar with these kinds of agreements, and they get negotiated every day.

When there is a party that is expropriating an easement over your property, there is a very well established way of appraising this. What happens, in a nutshell, is this:

a. The lawyers look at the value of the property before the easement has been expropriated;
b. They look at the value of the property after the easement has been expropriated;
c. They identify whether there is a difference between the two;
d. Whatever the difference is, that is the value of the easement they have just expropriated.

We lawyers, in our uncanny ability to capture and articulate the blindingly obvious, call this the "before-and-after" method. In fact, the before-and-after method has been around, as I said, in this province and elsewhere for a very long time. It applies whenever an expropriating body acquires an easement or engages in what expropriation law calls, in one of the great oxymorons of the English language, Injurious Affection. I rank that up with friendly fire! This refers to what happens when your property is being devalued but there has not been an acquisition of the actual ownership of the property. So, as I say, this sort of thing has been around for a long time.

We now get around to the question of how we deal with this for receipt purposes. If somebody wants to donate a covenant or an easement over a property which has certain historical features, and wants to donate it to the Alberta Historical Resources Foundation, Government of Alberta, City of Calgary, Calgary Civic Trust, whomever, if they want to donate this, how do we appraise this thing?

> Well, if we appraise it the same way that easements and covenants have been appraised for a hundred years, we use the before-and-after method. This is where we get into the tricky part because Revenue Canada, now called the Canada Customs and Revenue Agency (CCRA), is of course the ultimate body that decides whether it is going to sit tight with the receipts that are being issued to these kinds of agreements, or whether it is going to disallow the receipt and lead us all into court.

Today I intend to present you with three propositions:

1. Yes, there is a method for appraising the donation of covenanting easements. That is the good news.
2. No, Canada Customs and Revenue Agency does not use this method. As a matter of fact, CCRA as of Wednesday of last week (6th October 2000) informed me that they use a method which essentially says that it does not matter how devalued the property may be as a result of the covenant. It does not matter that the owner may have signed away $100,000 worth of development rights on the property. Essentially, their position is that they are ready to appraise this gift at maybe a dollar.
3. As was said in "The Hitchhiker's Guide to the Galaxy" in large friendly letters, DON'T PANIC!

Let me go through these three points with you. The first is, what is the method? As I have pointed out, the question of easements and covenants, Injurious Affection, has been around in expropriation law for a very long time. The first time (that I know of) in Canada when this kind of agreement was applied to an historic building for its protection was in the mid-1960s in Montreal. As Jeremy Collins pointed out to you, in the province of Quebec the same kind of agreement does exist. It is called a servitude there, and yes, this has been used for the protection of historic buildings for about thirty-five years.

At roughly the same time in the United States, a major movement started to occur as people were using these kinds of agreements there as well, particularly in the ecological sphere. The Internal Revenue Service had, in fact, issued an opinion in the mid-1960s saying that on the basis of straight legal logic, they felt compelled to adopt the before-and-after method. That was then entrenched in legislation in 1976 and again in 1980.

A number of articles appeared in Canada as the whole easement and covenant issue gathered momentum in the 1970s and 1980s, more on the ecological front, I must admit, than on the heritage front. There are, however, various law journal articles out of Osgood Hall Law School and various other places that said one after another after another, "Yes, these should be subject to the before-and-after method because that is the only logical way to approach them."

This finally hit Revenue Canada. In July 1990, Revenue Canada issued a memorandum of opinion to an environmental group on Prince Edward Island outlining the exact way that you appraise gifts of restricted covenants. This is appended to the materials, which are going to be distributed to you, and as you will see, big surprise – it is the before-and-after method!

There we have it. A hundred some odd years of the before-and-after method, then something happened. I do not know what happened, and I do not want to cast any aspersions on Revenue Canada – I should say that I have had the pleasure of working with some people who were setting up a charitable organization for the relief of cancer in rural India and these were all Revenue Canada people who founded it, launched it, ran it, and donated to it. If anybody tells me that these people have ice water running in their veins, I am sorry, it is not true. I have immense respect for these people, but I should add that in an organization with thousands upon thousands of people working for it, it is statistically impossible for an organization of that size not to have the occasional cosmic idiot.

By 1996 all easements were deemed by Revenue Canada to have used unacceptable appraisal methods (I am quoting here). It was the department's view that only Fair Market Value appraisal techniques and not before-and-after techniques would be accepted. The value of an easement would be considered nominal, as they are not bought or sold. That, I should add, is a quotation directly from Environment Canada.

What are they getting at here? Somebody within the department lost the memorandum of opinion of 1990. As a matter of fact, until I faxed a copy of that memorandum back to them two weeks ago, they had lost all record that it had been issued. Having lost their original memorandum having to do with the before-and-after method, somebody hit on the idea that in the open market place we look at Fair Market Value as the ultimate guide to how we appraise property, or for that matter anything else.

What does the Fair Market Value test mean? It means we look at what the collective cluster of people offering something in the open marketplace are ready to agree on, with the collective cluster of prospective buyers in the open marketplace. So the logically correct way of appraising a covenant or easement would be to ask, "What is the cluster of 'offerors' of covenants and easements prepared to accept from the collective cluster of buyers (people who are in the market to buy conservation covenants and easements)?" That will tell you the fair market value of the covenant or easement.

Of course, if you do not have a collective cluster of people who are offering conservation covenants and easements on the open market, and you do not have a collective cluster of people who are in the business of buying conservation covenants and easements on the open market, then you do not have an open marketplace. If you do not have an open marketplace, you do not have market value. If you do not have market value, you have to rely on nominal value – in this case a buck. That is how the thinking process works.

Of course there is no cluster of people in Alberta who are falling over one another to buy conservation covenants and easements. There is no groundswell of people coming forward in Alberta saying, "I've got a conservation easement to sell – anybody want to buy it?" Since that is your dilemma in Alberta, the appraised value of the donation of such a thing would be a buck. If this makes sense to you, you are way ahead of me!

Now, this is of course distinct from the way other countries appraise these things, as we heard in the United States. It is also distinct from the way expropriation law in Alberta has worked for a hundred years. It is distinct from almost anything else that I know.

It is also distinct from other gifts. Let us remember that the *Income Tax Act* of Canada recognizes several different categories of gifts. As a matter of fact, anybody who believes that Canada's two official languages are English and French has never read the *Income Tax Act*. We have charitable gifts, we have cultural gifts, and we have ecological gifts. In point of fact, they all rely on different techniques for doing all the calculations. Do not mistake any one category from another – that is totally fatal and it will give your chartered accountant apoplexy. I know one particular altruistic gentleman with a certain nature trust in Ontario who actually paid $20,000 in professional fees to figure out how to give his property away. So this is not funny.

In the case of ecological gifts, for example, you have a specific category of calculation that deals with capital gains in a certain way. Oh yes, I have to mention capital gains. Every time property appreciates in value and you decide to sell it, unless it is your personal residence, which is exempt from capital gains, the government wants its cut. It wants to have its share of the profit margin on the disposition of the land. That includes dispositions of the land via philanthropy. So if you give your property away, it can land you with a capital gains liability.

In order to calculate capital gains you have to calculate the value at the time of disposition, and you also have to calculate the value at the time of acquisition. You even have capital gains on covenants and easements. One thing that I have to say for these nice people at the tax office is they are consistent! So that means that if you are donating a covenant or easement, and if you want to say that it has a value (even if the value is only a dollar) at the time of disposition, you also have to figure out what value it had at the time of its acquisition: its adjusted cost base. So far, I do not know of a single accountant anywhere in Canada who has been able to figure that one out.

When the ecological community discovered what was happening in the 1990s (in 1996 to be precise), it realized that all the covenants and easements they were trying to negotiate on wetlands and forest and eskers and moraines and all those important parts of our natural heritage were being endangered because of this particular tax view now adopted at CCRA. So instead of trying to negotiate some kind of new deal directly with the people who were reaching this decision at CCRA, they said, "Forget this! We will go straight to the Minister of Finance and we'll get the Act changed." And they did.

In 1997, there was an entirely new process adopted by the Federal Government in order to deal not only with donations of land, but also with donations of easements and covenants. It had a revolutionary new way – it must be new because the budget papers said it was – of appraising donations of covenants and easements. It was the before-and-after method.

So where does that leave the rest of us who are not donating covenants and easements over wetlands – we may be donating covenants and easements over property that is perhaps a heritage property. It is not only if you are donating a covenant or easement over farmland. It is anything that is not ecological. What the Feds had done in 1997 was to say that "Yes, we will adopt the before-and-after method for covenants and easements, but only in the case of ecological gifts." So as I said before, ecological gifts are a completely different set of calculations from other kinds of charitable gifts or cultural gifts.

Where does that leave us? Well the answer came at four o'clock last Wednesday. I received a telephone call from the Department, now CCRA, saying, "Well, since the before-and-after system is reserved for ecological gifts, and you are not talking about an ecological gift, you are talking about an historic ranch house or something like that, it does not apply to you. Therefore you are back to the Fair Market Value System. Therefore you are going to have to be confined to a receipt worth about a buck." So that is where it stands as of four o'clock last Wednesday. What do we do next? Well the answer is, as I've recommended, is first of all DON'T PANIC!

1. The first thing that could conceivably be done is: try it and see. There is an eminent Calgarian by the name of Henry Zimmer who once postulated what he called the no-worse-off rule. If you adopt an interpretation of the *Income Tax Act*, which is favourable to your position, and if Revenue Canada says, "No," if you are no worse off for having tried it, why not try it? In other words, if you adopt the no-worse-off rule, you prepare an agreement for an easement or covenant in Alberta, you get it appraised by the before-and-after method, you have the appraisal properly done and then you get a proper-looking receipt, and you see what happens next.

That is certainly a course of action, which I have seen some people consider in other provinces. It is dangerous. There is a distinct risk to this. There are some people who think that maybe it could be done if you were dealing, for example, through a foundation like the Ontario Heritage Foundation, because after all the OHF is an extension or agent of the Crown in Right of Ontario. Maybe, according to some people, the CCRA might be a little less inclined to tell the OHF that they don't know how to appraise, than they would be to tell the rest of us.

> 2. Another possibility is to say, "To hell with it, we sue! We are going to issue our receipt. We are going to do everything according to the before-and-after method. We are going to have everything professionally appraised and let the chips fall where they may. We are going to establish a War Chest and we are going to fight. Do a test case."

All well and good, but the Department of Justice in Ottawa has more lawyers than you do. So it depends how risk averse you are. There are some people who say, "Let us take a public stand and shame them into adopting our position!" OK. Possible, again, but these people do not shame easily. I should say that I was in one meeting with the head of a major industrial organization and he was complaining bitterly about the chaos, which a particular tax rule was causing, in his industry. The representative, a very senior person within the Department, said, "Yes, but the difference between you and me, sir, is that I am governed by logic and you are not."

The position taken by the Department, and actually I have some sympathy for this, is that they do not bend to public pressure. They will only bend to a better argument, which leads to the proposition that:

> 3. Maybe you should negotiate on the basis of a better argument.

It is very difficult to negotiate with these people, however, because their standard methodology is that they like you to put your case in writing to them. Then they go off for a few months, and you do not hear from them for a while. By the time the answer comes back, it is engraved in stone. It is relatively difficult to have face-to-face meetings where you exchange ideas.

You may have a better chance if you are acting under the patronage of a party that is tremendously influential: a provincial foundation, a provincial minister, something like that.

> 4. It is possible to recast the argument. So far, in all the texts that have been produced everybody has assumed that the before-and-after method is so completely different from the fair-market-value method that the two will never meet.

Let us suppose the argument goes like this: you want the fair market value established on the going rate for covenants and easements. You think, at present, there is no going rate for covenants and easements. Well, in point of fact, you are wrong. There is a going rate for covenants and easements. The going rate for covenants and easements in the year 2000 is the very same going rate that has existed for the last ninety-five years that the province has been in existence. It is the going rate established by the various expropriating bodies that have written the book in Alberta on how you establish the going rate for these agreements. If the book says the going rate is the before-and-after method, and if the province does have jurisdiction over property, and if it does have legally constituted boards all across the province that make it their business day-in-day-out to establish the going rate for easements (whether they are for pipelines, hydro-electric lines or telephone wires) then that is the going rate. The going rate dictates a value of x and that is the value that should be attached to the receipt. That is the argument.

Has the argument ever been used? Not to my knowledge. Is it worth a try? Possibly. Maybe in the form of a direct overture, maybe in the form of a well-placed and well-circulated law journal article, I do not know. It is certainly something you could give some thought to.

> 5. Another option: do the same thing the ecological community did. Forget CCRA. Go straight to the Minister of Finance and get an addendum to the *Income Tax Act*. If in 1997, our good friends in the ecological movement were able to get the before-and-after method re-instituted for the appraisal of covenants and easements, maybe we can go there and say, "Me too, me too." Let us put ourselves in the same category. As Doug Franklin is undoubtedly going

to mention to you, there is already a very important agenda which people like the Heritage Canada Foundation are bringing forward on the tax front. Adding another item to the agenda may be exactly what the doctor ordered.

6. Last possibility: do any or all of the above. Use them in combination. Use them tactically. I cannot recommend any particular course of action at this point. What I can tell you is that if you do not – if you let this issue simply sit – then any major, broad-based initiative to recruit covenants and easements is almost certainly doomed to fail. You will not be able to sit across the table from a prospective landowner and say, "If you part with $100,000 worth of your development rights, you are divesting yourself, your family and your heirs of a fair chunk of your assets, but at least you have the consolation of getting a tax receipt at the end." A tax receipt of one dollar, Ladies and Gentlemen, does not cut it.

So if you do not do anything about this, you are going to be in very serious difficulty. But if you do, if you approach it tactically, if you do not panic, then I suspect that with the kind of blueprint that you have gotten today on how these kinds of programs can be set up, I think that you are going to be able to launch a very successful initiative and comprehensive program for covenants and easements in Alberta.

Heritage Canada Policy: The Canadian Tax Context
Doug Franklin, Heritage Canada Foundation

I would like to give you a quick overview of Heritage Canada's brief within the Canadian tax context. Marc Denhez gave an eloquent presentation on that Byzantine world of revenue, tax policy and Finance Canada. We have been working on this brief for some twenty-seven years, and although the going does not get any easier, I think we can see light at the end of the tunnel.

Covenants are highly desirable as instruments of preservation, combining protection and economic incentives. Easements and covenants for the protection and preservation of heritage buildings are under utilized in Canada. It is worth nothing that even in the United States where easements with tax incentives have been used for more than twenty years, there is insufficient case law relating to income tax, according to the U.S. National Trust. Although common law easements and covenants exist, it is highly desirable for all Canadian provincial and territorial jurisdictions to have these instruments embedded in heritage legislation, particularly if we are seeking a new federal income tax measure for donating them. We have heard from a number of sources that there is a strong distinction between common law covenants and easements and those embedded in legislation. One of the reasons, of course, why it is important to have easements and covenants embedded in legislation is that the legislation will supersede, in almost every case, the common law and the common law interpretations. Moreover, the narrower the definitions are in a legal context, the more useful it is for everyone to share the same level of knowledge and communication. Valuation is also made simpler.

So our organization, Heritage Canada Foundation, applauds what all of the provinces and territories have been doing in this area. Ontario is arguably one of the most successful jurisdictions in dealing with easements. We commend this province, and we believe that these measures will become even more important as we move the whole brief forward for federal tax relief in the area of heritage preservation.

I would like to say a quick word about our experience in working with the federal government. We have had frank discussions, we have had off-the-cuff exchanges, and we have had official missives. Until very recently, the government of Canada believed that any tinkering or tampering, as they saw it, with the official *Tax Act* (apart from trying to make it neutral) was an aberration. They see the Canadian tax system as eminently fair. Recently, there has been a move towards deregulation and towards flat taxes. In actual experience, just the opposite has been the case. A very good example, that we will talk more about in a moment, is the tax treatment of ecologically sensitive lands.

The premise of having public policy objectives as part of the *Tax Act* was stated very succinctly in the federal budget speech of February 28th 2000. I will read it briefly, because it is very important:

> The measures with respect to ecologically sensitive land are examples of how the tax system can be used to support the government's overall environmental policy. Other examples include accelerated depreciation for energy conservation equipment and tax relief for ethanol and methanol used in gasoline-blended fuels.

There, in clear print, is a specific part of that logic laid out before us by the government itself. In other words, it is no longer an aberration to think about the *Tax Act* as an instrument of public policy, but in fact as a very important means (given the level of taxation in this country) to help shape policy in desirable directions to benefit all Canadians. And I underline that – all Canadians. If somebody in Finance or the Customs and Revenue Agency even utters the expression "heritage buffs" we respond. We argue that this is for the benefit of all Canadians. Philanthropy, as it relates to heritage conservation, needs every means of encouragement.

The recognition by the Government of Canada that the tax system can be an instrument of public policy is a very important development and we are going to be watching it very closely. I will talk later about our current initiative as of this month, in this area. I want to probe a bit more, however, into the recent budget because we believe that it is the step beyond "Don't Panic!" It is, in a way, a blueprint for us. The proposed amendment to the *Tax Act* comes by way of a notice of a Ways-and-Means motion to the federal budget under donations of ecological gifts. It states that:

a) the provisions of the *Income Tax Act* relating to ecological gifts be modified to halve the income inclusion rate for capital gains from such gifts, other than gifts made to a private foundation made after February 27th 2000;
b) it requires the donor to file with the return of income for the taxation year in which the gift was made, a document obtained from the Minister of the Environment certifying for the purposes of the Act relating to charitable gifts, the fair market value of the gift as determined by that Minister.
c) it provides a donor with a right to appeal to the Tax Court of Canada, a re-determination by that Minister of the fair market value of a gift that has been made and provide that such a valuation apply for all purposes of the Act relating to charitable gifts for the two year period following the time of the last determination or re-determination of the value.

To us, this represents genuine progress in the tax treatment of gifts of heritage property. Marc Denhez's recent news is disturbing but will the Tax Court of Canada seriously look at a case in which someone has even offered a gift of land for the consideration of one dollar? Tax court cases over the value of gifts in this country continue.

Valuation is the key to the enterprise as far as the donor giving up a consideration, or giving up something without strings attached. This, of course, is for the public good. Such is a gift or a donation. Again, in the Federal Budget of this year, we have a comment, an enshrining of that measure in 1997. Normally, the value of a donated property is determined to be the price that a purchaser would pay for the property on the open market. As there is no established market for covenants, easements and servitudes, the fair market value of such restrictions on land use is difficult to determine. To provide greater certainty in making these valuations, the 1997 Budget introduced a measure to deem the value of these gifts not less than the resulting decrease in the value of the land. This is a very important principle, and I hope that it will be further refined.

I would like, for a moment, to talk about valuation and the process of easements and covenants in the United States, just for comparison purposes. I want to focus on valuation. Of course, we have our own case law in Canada. But occasionally members of the judiciary look at what are called informal sources of law. They look at the United Kingdom and they look at the United States. In the United States, as some of our speakers have already mentioned, the donation of easements on heritage property were made a permanent part of the U.S. tax code in 1980. But according to the National Trust for Historic Preservation, the most serious problem is the lack of a clear method for easement valuation that will make the tax benefits of an easement donation easily understood and simple to calculate. The valuation method in the United States considers three factors:

1. The first is market data, meaning the current market value of a particular building. We realize that this is not an esoteric exercise. It is what the real estate industry deals with daily. The data on three-storey warehouses in a particular area of town must be compared. If there is one next to it then what is it worth? What is the one on the other side worth? When was the last time it was put to market? What were the buyers willing to pay for it? So that is quite straightforward.
2. The second is the cost, based on the existing building but not the speculative factor. So that is what the buyer is willing to pay for a property "as is."
3. With the third factor, we get into income based on potential income and potential losses, for example, that a developer might incur by preserving a building in situ. Typically in the U.S. experience there have been decreases in the jurisdictions that use easements and covenants most widely. I thought we would mention several today. The Historic Georgetown Foundation in Washington, DC,

has found a decrease of 5 to 20 per cent. The Foundation for San Francisco's Architectural Heritage, which has over a hundred easements in that city alone, finds the range a 10 to 20 per cent decrease. A 50 per cent decrease of value throughout the United States is considered rare. One example of that could be where there is an historic building on a piece of property (a significant part of which is vacant) that could be developed and put to the highest and best use. With the imposition of restrictions on the development because of the historic fabric of the building, however, the developer is not able to realize the full potential. Therefore, there would be a great decrease in the value.

One particular case in the United States is worth citing: the so-called Thayer decision. It is related to both an historic and scenic property in Virginia, consisting of a farm, a principal building, outbuildings, and agricultural infrastructure. This particular case did get to the Tax Court and the petitioners, the couple known as the Thayers, claimed 42.6 per cent as the decrease of their property value. The Internal Revenue Service (IRS) claimed 20 per cent and the Court allowed 33 per cent. So you see it was almost up the middle in that case. It was interesting because when the IRS looked at the property, their valuation actually increased the value of some of the buildings, and decreased the value of others initially. Some buildings, such as utilitarian outbuildings at first glance did not have a great deal of value, but in the final analysis, the IRS actually increased the value of all of the buildings, because once the covenant was in place, no one could demolish them. They were not recognized as heritage properties as such, but once the covenant was put in place, that instrument protected them as if they were very important, as indeed they were, as part of the historic and scenic fabric of that particular ensemble.

What we are trying to do now, and have been working on diligently particularly in the last five years, is to get the entire tax system in Canada changed vis-à-vis historic properties. As a matter of public policy, we would like the owners of revenue-producing commercial buildings to be able to have choices about how they write off expenditures that would restore the heritage character of buildings. Until quite recently, we did not feel as though we had made much progress. In the Federal Budget in February 2000, however, the Minister of Finance unveiled a process, which happily is now well underway, of looking at the tax law and the treatment of heritage property. I believe that the path has been cleared for us, particularly in this very significant statement relating to ecologically sensitive land found in that Federal Budget. This strategy emphasizes providing assistance to encourage Canadians to take voluntary action to protect species and to make responsible stewardship an easy choice.

To Heritage Canada, it is a logical step to address heritage property. The government wishes to see more voluntary action. The government also knows it cannot afford to acquire all these places itself, or protect species, or protect heritage buildings (an endangered species). The Minister of Canadian Heritage, the Honourable Sheila Copps, said a year ago, "In the last 30 years we have lost 20%, or 1/5 of our heritage building stock." That is an endangered species, no less than organic forms that we cherish.

Their words: "To make responsible stewardship an easy choice." Well, if a chartered accountant who deals with the *Tax Act* daily cannot figure it out, Mr. and Mrs. John Q. Public certainly cannot. If they cannot do it, they will walk away. They will give their donations to the ballet or they will give their property to a university. I will read you the statement that the Minister made last February:

> The government is committed to the development of initiatives in support of the restoration and preservation of Canada's built heritage. Canadian Heritage officials (that's the Department) have undertaken discussions with provincial, territorial and municipal officials with a view to establishing a national register and conservation standards in respect of heritage property. These tools will be instrumental in assessing the necessity of financial support to sustain and ensure the preservation of Canada's built heritage."

That is a first in our twenty-seven years of advocacy in this area – to hear these words uttered by a Minister of Finance in a Federal Budget. In our discussions with Finance in the last few years, we have asked that immoveable cultural property be treated in the same way as moveable cultural property and ecologically sensitive lands. That logic now seems to have taken hold.

That is where we are today and many of you are taking part in this initiative within your various provincial and territorial agencies. We are delighted, and we applaud that initiative. I will read you part of our current brief to the Standing Committee on Finances:

> While discussions on a national register and conservation standards are now taking place, the Heritage Canada Foundation would like to seek the support of the Minister of Finance for the tax treatment of donations of built heritage property, easements and covenants, parallel to the treatment announced in the 2000 Budget for ecologically sensitive lands.
>
> The Foundation applauds the measures announced in the 2000 Budget that will encourage the stewardship of ecologically sensitive lands, and urges that similar treatment be given to those civic-minded Canadians who wish to make gifts of heritage buildings as well as easements and covenants. The Heritage Canada Foundation believes that such measures would prompt action on the part of heritage property owners wishing to donate property to eligible bodies, especially municipal heritage organizations.
>
> In making this recommendation the Heritage Canada Foundation believes that the precedent of measures for the donation of ecologically sensitive lands announced in the 2000 Budget should be applied without delay to built heritage, even as discussions continue on the consultations pertaining to heritage buildings announced in the 2000 Budget by the Minister of Finance.

So our reception in future appearances before the Standing Committee on Finance will be very revealing. One can hope that the Department of Finance will see the urgency of this. This conference, and the proceedings and findings of this workshop, will be critically important as we work with all of you to bring this issue forward and get some action on the national level dealing with tax and the treatment of heritage properties.

Calculating the Market Value: Appraisal Methods

Jason Ness, Calgary Civic Trust

Covenanting is one tool that can be used to aid in the preservation and conservation of our cultural heritage resources. In order to determine the value of such covenants for purposes of donation and tax relief, an appraisal of the property must be undertaken. This chapter will look at the method by which this appraisal is done. I will examine the way in which cultural property covenants are appraised compared with those of ecologically sensitive lands.

The Alberta Land Trust has described real estate as being a "bundle of property rights."[8] On urban properties these can include the right to:

a. subdivide;
b. construct buildings;
c. modify those buildings;
d. modify the landscape.

On rural properties additional rights may include:

a. restriction of access;
b. the right to graze animals;
c. the right to develop timber and mineral resources.

These individual rights may be separated from the bundle and transferred to another body in the form of an easement or covenant. Thus a covenant can be placed upon an entire structure, the façade, or on both the building and its associated landscaping.

If a cultural property is given outright (fee simple) to an accepted body, such as a municipality, province or charitable group, it will be appraised on the basis of its Fair Market Value (FMV).[9] The definition of FMV, as outlined in the "Canadian Standards of Professional Appraisal Practice," is as follows:

Market value means the most probable price, which a property should bring in a competitive and open market under all conditions requisite to a fair sale, the buyer and seller each acting prudently and knowledgably, and assuming the price is not affected by undue stimulus. Implicit in this definition is the consummation of a sale as of a specific date and the passing of title from seller to buyer whereby:

a. Buyer and seller are typically motivated;
b. Both parties are well informed or well advised, and acting in what they consider their best interests;
c. A reasonable time is allowed for exposure in the open market;
d. Payment is made in terms of cash in Canadian dollars or in terms of financial arrangements comparable thereto;
e. The price represents the normal consideration for the property sold unaffected by special or creative financing or sales concessions granted by anyone associated with the sale.[10]

When a donor gives a covenant, which is a giving up of certain rights over their property but retention of its use, the FMV is not applied to the property itself, but rather to the covenant. As covenants are not a tradable commodity, the FMV is virtually nothing, and therefore no tax advantage can be accrued.

It is worth looking at the way in which ecological covenants are appraised as there is similarity in intent, and because ecologically sensitive lands are valued in a radically different and beneficial way. For ecological covenanting, appraisals are undertaken within the Ecological Gifts Program. Similarly to cultural property, eco-gifts can be either outright donations of ecologically sensitive land to a government or an approved charity, or a partial donation by way of a covenant that would restrict certain rights over the property.[11]

The Ecological Gifts Program is governed under the *Income Tax Act* of Canada, but is administered by Environment Canada. The appraisal process is overseen by the Appraisal Review Panel, which consists of a Chair, five regional appraisers, six additional appraisers, a planner, and a legal specialist.[12]

The appraisal process involves three steps:

a. certification of the property's eco-sensitivity;
b. designation of a qualified recipient agency;
c. determination of the fair market value.

The determination of the property's value is undertaken by an independent appraiser, who must be a certified member of the Appraisal Institute of Canada for valuations greater than $25,000. For complex or high value properties two appraisals are recommended. The Appraisal Review Panel then assesses the valuation, with costs born by the federal government. They can either accept the original amount, or put forward a revision. If the donor does not agree with the panel's final determination, they may apply for a redetermination.[13]

These appraisals are done strictly on the basis of property value; the ecological value of an eco-gift is only taken into consideration when determining if the property qualifies for the programme, it does not count in any way towards the determination of the market value. The appraisals are determined according to the Canadian Standards of Professional Appraisal Practice.[14]

As of 1996, the process of appraisal used the before-and-after method. An appraisal is undertaken determining the value of the property as it currently exists, then a second appraisal values the property as it would be with an easement or covenant in place.[15] This is the major difference between cultural and ecological covenants:

a. cultural covenants are assessed only on the value of the covenant itself, for which there is no market;
c. ecological covenants receive the value of the development potential that has been lost.

As recently as October 2000, the Canadian Customs and Revenue Agency ruled that ecological and cultural gifts are completely different and therefore it remains appropriate to assess them differently for tax purposes.

This potential is determined through the principle of Highest-and-Best-Use, which takes into account both the negative and positive aspects of the property and its surrounding area as to the highest potential value of its development.

Negative factors could include:

a. proximity to other encumbered lands;
b. a negative attitude towards development of ecologically sensitive land by the local authority.

Positive aspects could be:

a. river or lake frontage;
b. viewscapes;
c. potential for intensive subdivision.

It is normally a change in Highest-and-Best-Use that will be most dramatically affected by a covenant, and thus will determine its value.[16]

Ecological easements are relatively new to Canada, so appraisals must be done on a case- by-case basis. Eventually, lists will be compiled that will enable assessors to compare the values of actual encumbered and unencumbered properties. Without such precedents, the determination process is ambiguous and relies largely upon the appraiser's logic, reason and experience.

Some tools that can be used are:

a. determination of the net-income generated;
b. comparison of re-sales of encumbered properties in the United States, where they have existed much longer;
c. comparison of properties with similar potential.

In February 2000, the federal Budget reduced the taxable portion of capital gains associated with ecological gifts (including covenants) to 25 per cent,[17] thus creating a healthy incentive for donation. No such incentive currently exists at the federal level for the donation of culturally significant properties, thereby impeding the effectiveness of cultural covenanting. To this end, groups such as the Heritage Canada Foundation and the Calgary Civic Trust are lobbying to bring cultural gifting in line with the appraisal method and tax benefits received by the ecological community. With such a change, covenanting could become a powerful yet flexible tool for the conservation of our cultural resources.

SECTION 4

Discussion

This section pulls together the presentations given over the two days of the Covenanting Workshop. Prof. Michael McMordie reviews some of the key points of Time, Money the Process and Value in covenanting Heritage Property. Frits Pannekoek then introduces and chairs the afternoon development of Covenants for a number of local case study buildings provided by individual property owners. In this session workshop participants use local case studies of heritage buildings to begin discussion toward the development of individual covenants. A panel discussion follows assessing the impact of covenanting on development.

A Review of Key Issues and Design of a Covenant
Michael McMordie

A Review:

Time: there is no quick fix to the issue of time. We have to look at the way the development industry has to look at it. As Harold Milavsky makes very clear in his discussion of Banker's Hall, we have to consider the implications at least twenty years from the completion of the project. Clearly, time is a dimension that is enormously important and time does not flow uniformly in this process. The circumstances are always changing, and what is possible one year may not be possible the next year and may become possible again the following year. We need to have ways and means of responding to these situations, just as the developer and development industry do. From the political perspective, we must be aware of the changing moods and memberships of city councils. All of this has to be figured into the mix.

Money is central to some of our discussion. Marc Denhez made it very clear how crucial, complex and uncertain the tax situation is for preservation. Covenanting, if it is going to become the kind of tool that we want it to become, depends to a considerable degree on the favourable ruling on evaluation. The viability of many of these projects also depends upon very carefully balanced budgets, with very narrow margins. That is both a problem and an opportunity, because if we can establish historic covenanting as a mechanism that attracts significant valuations and therefore enables significant tax receipts to be given, it becomes a significant lever in the whole process of financing of some of these projects.

Process is important and that process has got to involve all parties in extended negotiation. This is normal between the development industry and the City in commercial development. The example of the Convention Centre, Hyatt Regency Hotel block is, I think, a very favourable omen. The fact that we could bring together the City, the developers, the lenders, and the heritage interest (as represented by Rob Graham) in the heritage planning process and produce a significant new project is greatly important to the city in its economic and cultural development. But, maybe even more significant because it integrated and interwove both the new and the old. That kind of model is one that we need to continue to look for if we need to accustom people to thinking in terms of extended and careful negotiation, which involves and respects all the pertinent interests in order to produce results that are satisfactory for all the parties involved.

Values, finally, of course money is one measure of value, but we are here because we are concerned with other less quantifiable values, such as heritage and cultural values. The discussions made clear that while these values are present in the community and latent, they needed to be mobilized and brought to the fore. The success of the heritage district, to which Rob Graham referred, and the revitalization of Stephen Avenue is based on people's realization that the environment that we are working for is attractive, valuable, enriching, one to be supported, one to be enjoyed, one to be used. That means that as we go forward, we have a responsibility, I think, to a public education process. I know a number of you have been working at this quietly over the years, but we have got to find more effective ways of bringing to the attention of the people of this city (I am speaking as a Calgarian) the virtues of living environments, shopping environments and working environments that respect the old and see it as a necessary ingredient of the new as we build for the future.

Those are the conclusions that I arrived at, and I am sure much more will percolate to the surface over the coming days and weeks as I reflect on what we have done here. I hope that is a useful if over-concise summary of some of the things that have been said here.

A number of projects based on these values are emerging. Harold Milavsky, for instance, pointed to the fact that even municipal designation (with what has been seen as the threat of compensation) has never been properly tested. I think one of the projects might be to do some case studies on what that compensation might amount to and we might be surprised by what we learn. We are beginning that process now in this workshop with some hypothetical examinations of real buildings.

Designing a Covenant: Frits Pannekoek

The development of a model covenant that could be applied to real buildings has been challenging. First of all, we would very much like to thank the people who volunteered their buildings. To expose ones self to a group and their queries requires great generosity.

The process we are going to follow in the exercise is to work with owners and attempt to draft a covenant. A draft or model covenant borrowed from the United States and with some changes is included with the workshop materials. The assumption is that this document is a model for Alberta. That doesn't mean it cannot be changed, but it is something to work from. Another assumption is that the Civic Trust has been designated as a covenanting agent under Section 29 of the *Alberta Historical Resources Act*. (The Calgary Civic Trust was designated an agent in 2002).

The last assumption is that when you sign for any of these properties, you are to pretend you are members of the Calgary Civic Trust and you are working with this owner to come up with some reasoned situation that might be in the form of a covenant. At the end of the discussion you may conclude that a covenant will not be viable. There may be another more appropriate solution. So simply take what we have learned in the last few days and apply this information to a given situation. Covenants at the best of times and the worst of times are extraordinarily complex, so our expert guests and presenters featured earlier in this volume made themselves available for questions during the workshop negotiations. Five unique heritage properties were submitted as case studies for the workshop. The following is a description of the property by the owner or designated person and a summary of the workshop discussions.

Naismith House, Calgary, Alberta

The Naismith House

1119 Sydenham Road S.W.
Calgary, Alberta
Presented by Mrs. Lucile Edwards (Owner)

The Naismith House in Mount Royal is on a very charming little street called Sydenham Road, which is only two blocks long. The property was purchased from the Canadian Pacific Railway in 1910 and the carriage house was constructed the same year. The following year the house was constructed. It is currently a Provincial Registered resource. This designation does protect it to some extent, but if the house was to be sold, I am not totally convinced that if a new owner wanted to tear it down, the province would protect it. I would hate to see it go. So I am interested in getting a covenant on this property.

The carriage house is one of the few left in Calgary and may be the last left in Mount Royal. The carriage house is in its original condition, but an addition was made to the house, preserving the integrity of the original building. The interior is largely original as well, at least in the most interesting areas of the house: the living/dining room. Modernization has been done and there probably needs to be some more as time goes on. For the most part, I think the character of the house has been maintained, with lovely woodwork on the inside and a very grand entrance. A very nice yard slopes up to the house with large clumps of spruce trees that are terrifically hard to keep trimmed into the original round shapes.

Discussion of the Naismith House

Presented by Jason Ness

Our group is presenting the Naismith house and we are pleased to have the owner of the house in our group as well. The first thing we had to look at was the fact that the owner was also a board member of the Calgary Civic Trust. We wanted to establish whether there would be a conflict of interest or a perception of a conflict of interest, and what that would mean. Would the board member have to step down or stand aside from the process of dealing with her own property? That is something that the Board of Trustees will have to decide.

The next thing we looked at was whether or not a covenant is a deterrent to the sale, as this is important to the owner. Questions like, "Will be there a commitment to maintenance in the future?" and "What provisions would be made in terms of future insurance on the property?" "Is that a deterrent to future ownership, if and when it changes hands?" We discussed what other improvements would be made to the property and perhaps doing a collateral agreement. Our decision was that the provisions are already in the covenant. The discussion then looked at what kind of public accessibility would be adequate. We decided two days per year would be acceptable to allow for house tours.

Another issue that was discussed was the perception of banks towards these covenants. Would it be a problem with a mortgage? We decided there was a need to educate banks on what exactly covenanting means and also that the owner needs to be very clear on the understanding of the legal context.

We decided that covenanting provides a very comprehensive way to preserve heritage properties because it defines the specifics. This property is special in that it is designated as a provincial registered heritage resource, which provides grants for restoration and is at least a temporary safeguard towards preventing improper changes or demolition. But, we also realized that registration is impermanent and can be changed due to political will, whereas covenanting, because of its legal nature would be in place in perpetuity.

Next we talked about valuation and we decided we lacked the expertise to actually recommend an amount, but used some of the U.S. figures of 15 to 20 per cent on average.

We went through the property and listed some of the features that we decided were important things that need to be preserved. We do recognize that for a formal covenant we would want a qualified architectural historian to do a baseline survey for the property, so this is our amateur listing.

The features to be preserved are the entrance space and the fabrics within, the second floor front rooms and master bedrooms, for their interiors. The exterior and the carriage house should also be preserved. We recognized that there was a definite threat of subdivision and a separate residence being built. We wanted to prevent that from happening. There are also some significant features in the garden, such as mature trees, a prominent walkway and things that are typical of the Mount Royal community. We decided we would allow no changes unless there was approval from the Board of Trustees. We would use the provincial guidelines to govern any changes. We also talked about prohibition on changes of use that were unsympathetic, without approval.

We talked about commemorative plaques, both a Provincial Registered Resource designation plaque (which is in place) and the possibility of a more informational type plaque. We decided there would have to be an agreement between the Trust and the owner as to what type of plaque would be erected and where, because the owner may or may not want people coming up to the house or may or may not want people stopping their cars out front. So we decided there would be plaques, but there would have to be consultation and agreement.

One of the clauses in the covenant covers what would happen if the Calgary Civic Trust could not take on this responsibility. We decided that two other organizations, the first being the Alberta Historical Resources Foundation and the second Heritage Canada could become possible holders of a covenant.

Question: What you have presented gives a pretty clear idea of what you don't want to see happen at the property. I am not sure that I have grasped yet exactly what you do want to see happen at the property and whether you are making the path as easy as possible for that objective to be achieved. Is there a particular brand or style of work that you see to be desirable and is your agreement being structured to facilitate that? What kind of future do you really want to see there?

Jason Ness: I don't think we need any physical changes. The house is in very good shape. That is why we didn't specifically discuss that.

Question: You mentioned that the features worthy of protection are both internal and external. Do you have a view as to the relative significance of interior and exterior?

Jason Ness: The exterior is probably more important, although there is something I didn't mention. There is an addition on it at the back and it is sympathetic. I will read the heritage character statement: " This intact house is a good example of the Tudor Revival style. It has half-timbered dormers with roughcast stucco, an open veranda with the entry door framed in a large arched window and walls clad with painted wood shingles. The original coach house remains intact at the rear. The expansive sloped front yard is characteristic of the Mount Royal district with its terracing and mature trees, curving walkway and side driveway. Two prominent Calgarians have owned the property prior to the current owner. Peter Naismith, the first owner, was head of the Alberta Railway Irrigation Company and Arthur Smith, the second owner, was a prominent lawyer and MP for Calgary West."

There is quite a large addition at the back, but it is sympathetic and we talked about whether or not there should be a partial easement that would not include that, but we decided that the covenant should include the addition as well because if not, it could be torn down and something less sympathetic may replace it.

Question: The reason I asked the question is that if the interior were a large part of the reason for us wishing to become involved, then given the substantial investment, which the board would be making in terms of monitoring and discussion, I wonder if enough significance has been attributed to the access. If the interior is a big component, are two days a year a fair reflection of that importance in terms of public being able to see it?

Jason Ness: There are quite significant features inside. There is a lot of fine woodwork and it hasn't been altered. It was in consultation with the owner that we came up with two days.

Question: Is there provision for that to change, if the interests of Calgary residents develop so that there is an upsurge of interest in visiting the interiors of old homes? Does the Trust have the ability to increase the number of days, or is it tied into what might be a rather restrictive arrangement?

Jason Ness: Some of the group are saying yes. I don't know how the owner feels about that. Could we set up a negotiating process within the covenant? Is that possible?

We did discuss the question of site signage and whether or not the Trust had the right to advertise this as part of a tour and a negotiation process was set up to deal with this. We discussed offsite signage (maps) that might inform people interested in the house.

Question: As a trustee, this is going to impose obligations on the Civic Trust that are going to cost us money. How much of an endowment is going to go along with these easements so we can recoup some of those costs?

Our practice on something like this is to calculate the cost of monitoring, which is sending a professional architect out once every two years, not every six months. If we were a local organization it is more likely to be once every six months. The cost, if there is a cost, is of staff time, which would be minimal in this case. More significant is what we would call an enforcement fund. The way that we calculate the enforcement costs, and this is purely arbitrary on our part, is a formula that we have come up with to calculate the cost of a lawyer's time for two hours a year assuming we are not likely to use an outside lawyer every year. But maybe every five years there will be an issue and that will give us ten hours of a lawyer's time every five years, calculated on the basis of the current going rate for a lawyer, which depends on the place you are. In Washington, D.C., it is probably close to between $400 and $500 an hour and most places it is more like $250 an hour and you amortize that and make it into an endowment by multiplying it by twenty.

So if you say $250 per year times twenty gives you $5,000. So we would need at least $5,000 as an endowment. Therefore we are probably looking at a charge to the owner for the privilege of donating this property to us of somewhere around $7,500.

Question: One related issue is the first one you brought up: the conflict of interest issue. I am curious to hear from my colleagues how they would anticipate dealing with that. I think that you can protect yourself from any suggestion of collusion. In this case I would advise an outside consultant to craft the easement. That would also come with a cost, and we would probably pass that cost onto the property owner.

Question: The one piece that is missing in this covenant, which I don't think would apply to any others, is that this house is already designated under the provincial statutes. So one of the things that the covenant would have to acknowledge in some fashion is that, the covenant conditions not withstanding, it does not derogate from the minister's authority to approve all prospective changes to the property. So regardless of the approvals given by the trust holding the covenant, that is still not enough permission to the landowner to proceed with alterations until they have the minister's approval.

Answer: Somewhere in the easement there is a standard clause that says that all other permitting and approval requirements must be followed in any event. That would cover it.

Question: What is the advantage of putting a covenant on this property if the owner has to spend $7,500 to do that and designate as well? Taxes?

Answer: I suppose, in theory, if "before-and-after" was consistent, there could be a tax benefit. Secondly, and I should let Marc answer this because he knows it better, the $7,500 is in fact a gift to the Trust and is tax deductible. One could argue that the designation is double dipping, but it makes one eligible for certain grants from the Historical Resources Foundation so, in effect, you are taking advantage of the tax system and creating an eligibility for repair to the heritage fabric. That is the balance.

Answer: There is also an element of certainty associated with covenants that you don't get with designation. The covenant will spell out in considerable detail the dos and don'ts. The designation is a legal requirement for referral in which the minister's discretion is applied absolutely. Generally, a landowner has only a rough idea of what is and is not permitted. Covenants tend to be far more explicit and go into some considerable detail so that there is a benefit of certainty relative to the property that comes with a covenant, rather than one that just comes with a designation.

This is a registered property, which means that after a certain number of days' notice, any subsequent owner could engage in any activity up to and including demolition, if it were permitted by the jurisdiction, which the covenant wouldn't allow.

Comment: Marc Denhez. As far as the question of the tax implications is concerned, the first part of it is what about the $7,500? Yes, there is no question it is receiptable. As far as the question of the donation of the covenant is concerned that is a little bit trickier. Even if we put aside the question of whether the before-and-after method is applied, let's assume that we succeed in enticing the Canada Customs and Revenue Agency to adopt a position more favourable to the before-and-after method. In the event that the property is already designated at the time of signing, the covenant is already subject to a regulatory regime, which has removed a fair chunk of the development potential of the property. If that is the case, then the taxman will ask the question, "Well what is the donor actually giving away?" If the taxman comes to the conclusion that the property is already so regulated that there are no development rights to give away at the time of signing the covenant, then the taxman could very well come to the conclusion that regardless of the before-and-after method, there has been no receiptable gift.

So given that situation, my recommendation would be that if at all possible, the covenant be entered before and not after the property has been subjected to a stringent regulatory regime.

Question: I wondered if that comment would also apply in a jurisdiction like Nova Scotia where there is no designation without the owner's consent, therefore the owner is effectively giving away rights on a designation.

Comment: Marc Denhez. The only time where this issue has come before the courts was a case that was decided by the Quebec Court of Appeal, in which the question was: "When there is a designation that has reduced the development potential of the property, does it make any difference that every time you apply, you get permission?" The answer given by the Quebec Court of Appeal was, "No, it makes absolutely no difference at all. " There was a situation where, yes, a property had been designated, but for the kind of development for which the property was appropriate there had never been a recorded incident of a development application having been refused. Ever. So the assessment authorities said, "Yes, it is subject to designation but it is still worth a lot of money, because in practical terms there is lots of development potential attached to the property, because all the owners do is apply and we know perfectly well the application is going to be approved." The answer given by the courts was that makes absolutely no difference. The owner still has to apply and until the application has been made and approved, the development cannot proceed.

The court will assess in light of what is feasible today under the existing status quo. It will not assess based upon a hypothetical application and a hypothetical approval. In that case, the value of the property for assessment purposes was reduced to one dollar. Given that, there is flip side to this coin. If there is, in fact, no development potential that would be recognized by the court as attaching to that property, the next question is if you then sign a covenant on top of that, have you given anything away? Well if you have the same judges that you had at the Quebec Court of Appeal, the answer would be no, you haven't given anything away.

Dr. Hughes' House

215 – 11th Ave N.E.
Calgary, Alberta
Presented by Allison Robertson

I am presenting for the owner who could not be here today, Kelly Smith. Dr. Hughes built the house in 1914. He was a prominent citizen at that time. Kelly is most concerned about preserving this house, but he is interested in possible moneys to help him do this. He is restoring and renovating the house on his own and he hopes to make it into a bed and breakfast. When he bought the house, it was basically derelict. The plumbing had burst and the walls were ruined. There is a lot of wood trim, mouldings and very old cabinet that the doctor had used, when part of the house was his infirmary. Kelly wants to add dormers on the third story of the house in the future. The house is two and a half stories.

Dr. Hughes' House, Calgary, Alberta

Discussion of Dr. Hughes House

Presented by D.W. von Kuster

We are dealing with a house on 315 – 11th Avenue NE. The owner, Kelly Smith, wants to turn the house into a bed and breakfast. Over the years, the house has been damaged by water, so I believe the interior has been gutted completely with the exception of an interior cabinet, which we have identified as being built into the wall in the dining room. The north, east and west facades are architecturally important for the area. The house is a two-and-a-half storey and as most of the surrounding houses are one storey, it is a landmark. We believe it is an interesting feature on the block itself.

We have taken the typical covenant and just added some things to the sections. The first section is 2.2. Prohibited Activities. What we will do in subsection G is to specify that the medicine cabinet must not be removed or altered. Presently it is painted. If the owner decides to fix it up in the future he should restore it to the original finish by sanding it down and determining its original appearance.

The east, west and north facades, which define the original character, must be preserved, including the windows, the veranda, and the siding.

The windows have been changed to modern nine-over-nine and six-over-six windows. We presume that the original windows were one-over-one and there are a couple still left in the structure, so we will attach a diagram to our covenant saying these ones remain and if the owner makes changes in the future, the windows have to be whatever style they were originally.

The handrails going up the steps to the veranda are presently steel, so we would like those to be removed if the owner restores the house. The exterior doors are modern and the base of the veranda has been parged (stuccoed), so we would like to have that removed as well.

The house has been recently painted. We don't know the original colour, but we would like to return the house to its original colour, or at least work out a sympathetic paint treatment.

The next section is the south facade that we deal with in the conditional rights. The south facade includes a dining room addition. We would allow the owner to build back there if he wants to, but the south facade should not be altered in any way that infringes on the design elements of the northeast and west facades of the house.

We spent a lot of time discussing possible dormers on the third floor. We decided in the end not to include them in the covenant at this time.

The Lorraine Apartment Building

12th Avenue S.W.
Calgary, Alberta
Presented by Sally Jennings

I am standing in for Neil Richardson, who is not able to be here. The "Lorraine" apartment building is interesting because it is only a burnt-out shell. It has very attractive brickwork, good textures, and good colours. It has three stories and some important terra cotta on the balconies in front.

Lorraine Apartment Building, Calgary, Alberta

It faces 12th Avenue S.W., facing south. It is a landmark building and very noticeable.

Three years ago, it was burned and all the historic fittings have been lost inside. Neil and his partners are recreating it and turning into an office building. I think the roof joists have just gone. The building takes up most of the footprint of the property, with little space on either side, but sufficient for angle parking there. The building will become offices, because they are more financially viable than apartments, although it is in an area where there are a lot of residential apartment buildings.

The building is turn of the century, 1912, when everything else went up in Calgary. At the moment, it has no use, because it is a building site, but Neil Richardson is a very concerned owner and wants to preserve its heritage nature.

Discussion of the Lorraine Apartment Building

Presented by Sally Jennings

The Lorraine building is the burnt-out shell. We have decided, as a body, to go to our lawyers tomorrow and make an offer on this place. We are going to have this as a category 'A' heritage building (on the municipal inventory) where we can get $75,000 every five years from the Alberta Historical Resources Foundation to maintain the facade. We are going to designate the building therefore, and because it is surrounded by at least fifteen-storey buildings, we are going to sell the air rights, so we will have no problem getting this building into good condition.

So with Jeremy's kind help, the basic idea is just to throw a net over the building – a general easement on all four facades, including the building as a whole, because in the plans it is now four storeys, but they envision building a fifth at the top, so we want to have control over the fifth floor being set back so it will not be visible from the street and the building will still maintain the same appearance, as well as any other provisions that might be changing. The basic principle is to keep the easement very simple, so that it can be monitored very easily. Ideally, when the people who put the easement on have gone, no one needs to go back to consult them on the original intention. So the simpler it is, the easier it is for everybody to understand and monitor and apply.

While we do that, there is going to be a secondary control in that the Civic Trust, or a heritage foundation, will work together with the owner in creating a master plan for the restoration and renovation of the building. The interior of the building is gutted, so there will be very big changes on the inside, and that can't be helped. So, as long as we have these two documents side by side (the master plan and its connection with the easement), we will ensure a good outcome when the architectural changes have been made.

In terms of monitoring, Paul was talking about how expensive it is get a lawyer to visit the property so we thought we would get a planner to walk by once a year (because it's the facade that we are really concerned about) as they are going to Singapore Sam's to get a take out, so the cost will be minimal. So I think we can quite adequately monitor the facade.

It is important to keep the principles manageable in the long term, so that in a hundred years' time when wood is $10,000 a foot we may want to change the provision that the windows have to be wood.

We haven't had the benefit of a tax assessor with regards to donating the façade easement to a heritage foundation, so we don't have a feeling for the before-and-after value of that facade. The Lorraine is standing, but derelict. In terms of valuation – the building was bought about a year ago for $400,000, and it is on a major street within a mile of downtown. There are big high-rise buildings around it, both offices and apartments, so it has a huge potential for development. The owner believes that when the building has been renovated it will be worth about $4 million.

Comment: Marc Denhez. Assuming we take the before-and-after method, my guess would be that the valuation would have to be approached by looking at the development potential of the property without the facade and looking at the development potential of the property with the obligation to retain and use the facade.

I would suspect that here in Calgary you would be successful in finding a number of eminently qualified appraisers who would tell you that to carry out a development without having to use the facade would save you thousands of dollars compared to the prospect of carrying out the development with the obligations of conserving and reusing the facade. Based upon that assessment, you could probably get quite an interesting figure for your before-and-after method of appraising that covenant.

Question: I don't know the building. Can you describe how the fire has left the exterior? The interior is clearly is gutted, but how much of the exterior is left and how much additional work is going to be needed during restoration to bring the facade up to the standard you want to see?

Answer: As far as I know, the building has not had much maintenance for a long time, so there would have to be some work done, such as re-pointing the bricks. I don't know what you do to terra cotta but what I can see from the ground it looks all right.

After the fire, the city sent an engineer to look at the structure of the shell to see if it was stable and it was except for the top floor where the fire started. That had to be reinforced in case it might fall away. Recently they have been putting new framework in the roof and tying the walls back together.

Question: Are you going to require restoration of the windows, or are you going to let them pop in vinyl-clad windows?

Answer: We thought we would restore all the windows on the south side facing the street and also on the east façade, which is open to a side yard. The west facade faces directly onto a high-rise building so there is no view of that facade. We could take windows out of the west façade and put them in the east. We could then put modern windows in the west facade. So we could preserve the woodwork on the south and east facades and let the west go, that is, in principle. If the owner wanted to restore all the windows, that would be good.

The owner mentioned that, effectively, the front facade, the entry way and the stair well are the only historic features that are left. The brick on the outside has suffered some smoke damage, but I think it is just a matter of repointing and cleaning. The terra cotta and the design would be maintained on the front and east, while the alley and west would allow for some change in the design.

Some of the windows are quite interesting because they have a bay affect, but in fact the building front is flat. They are not standard windows, and would have to be recreated.

Comment: I think it would be useful to explain that there is no roof on this building. It is just four walls and some of the walls are in extremely poor shape and that there are some parts missing. The Ontario Foundation's experience (in one instance where it took an easement on a building that had suffered severe fire damage) is that it is very difficult to place an easement on thin air – on a building that isn't all there. In those instances, we have required the owner to attach as-is plans of the building and projected plans of the restored building and have these attached. The easement covers the restored building as it is depicted on the plans that are attached.

How you get from the as-is plans to the projected plans would be the basis of a side agreement or master plan for the restoration. In that way, the easement would be the glue, while the master plan for rehabilitation would be the road map for how you would achieve the end result.

You can also apply for an upgrade for the baseline documentation so that there is a new baseline after the restoration, and that would become the baseline of the easement.

The Robertson House, Calgary, Alberta

The Robertson House

Dorchester Avenue S.W.
Calgary, Alberta
Presented by Nora Robertson (Owner)

I am a descendant of Don Robertson who bought this house in 1942. The builder and previous owner was the "hanging judge," also known as Judge Walsh. It is on the corner of 8th Street and Dorchester Avenue S.W., and I have always referred to it as my little cottage on the corner.

Unfortunately when my father died in 1992, we had a family auction and some of the pieces left the house. I have kept as much as I could, but some of other pieces went to brothers and sisters and have been dispersed. It is a beautiful house. It has a lovely entrance it is two and a half stories. Of the twenty-six rooms, seven are bedrooms, six bathrooms, a living room, dining room and two sun porches, a billiard room and a rumpus room in the basement plus all the usual basement things, fruit cellars and cool bins and washrooms. There are a lot of things that are still intact. There has not been too much that has been changed in the interior at all. It still is pretty much as it was when it was built. I have had it re-wired and re-plumbed, but other than that I haven't done anything to it.

It is a wonderful old family home, and I was raised there. I went away and got married and raised my children and came back to look after my father and I have spent the last twenty-three years there.

Discussion of the Robertson House

Presented by Fergus MacLaren

We were fortunate to have the owner of the house sitting at our table giving us her wisdom and experience.

I think that it is important, particularly when we are talking about a house, that both the legal and human aspects are considered.

The property is located in Mount Royal, where many houses are worth a lot of money and as a consequence there is a heavy tax burden. Several years ago at the MVA (which was the recent tax assessment), it was valued at $1,060,000. The tax for that was around $10,000. The heating bill alone is $800 a month. For someone such as Nora, living on a fixed income, the reality is that she could be forced out of the house. Her family has lived there since 1942, grew up in the house and has maintained it.

The house has twenty-six rooms and is 4,500 square feet. It is a substantial property for a single person. It is not designated despite the fact that her father was one of the city's civic builders and it is an important historic resource. Designation is therefore a possibility. So we started talking about designation. Marc's comment was to look at covenanting first and then go to designation afterwards.

Now because the property is large, it could be subdivided. If designation is considered, the property could not be developed. In that sense you are removing the value of that property with the development rights for it. So one of the things that we discussed with our legal experts was putting a covenant on the house and the carriage house and then designating them, allowing the property to remain separate from the covenanting of the house and the carriage house. The covenant would be simple but created to realise the potential of the house rather than its current state. The covenant would maintain the character and the integrity of the house, without discouraging potential buyers who might want to make some changes.

Secondly, once the covenant is in place, designation of the structure rather than the land could take place. The development rights of the property would be maintained, but the value would be enhanced because it could be subdivided in the future because the house itself is preserved by covenant. In essence, the intent is to allow the present owner to continue to live in the house. She has its best interest at heart. The intent is also to ensure that it is preserved despite a change of ownership, while maintaining its value and development potential.

Comment: Marc Denhez. What we were discussing here was that the designation process at least has certain tangible results in the immediate future. If you enter into a provincial scheme then there are certain very specific grant provisions that exist at the provincial level. There is a bird in the hand principle here. It is certainly attractive to enter into a designation scheme, because rather than wait until the cows come home for the tax man to render the decision we want on the subject of covenants, at least the designation approach has that appeal.

The difficulty is that is everything is designated and everything is subjected to a fairly stringent regulatory regime, which then drastically reduces any attractiveness that a later covenant might have from a tax receipt standpoint. So the question that we wrestled with was how to proceed given that reality. It would appear there is a possible avenue here and that is if you were to undertake an arrangement to have the property designated in the short term, the designation would apply to the entire property. The reason for that is simply a matter of labelling, because as soon as you have a designation, the designation is supposed to be registered on title and you have to name a property to which the designation applies, otherwise you can't walk in and deal with the Land Titles Office. So designation applies to the entirety of the lot.

The designation order itself, however, can restrict the regulatory regime to only part of the property, for example, to the four exterior walls and the interior, but the regulatory regime does not apply to the grounds. So it is possible to set up a designation that nominally applies to the whole lot, but in regulatory terms only applies to the building.

So if that is done, then the building becomes eligible for whatever grants the designation system triggers. That leaves the rest of the property, including the grounds, up for grabs in terms of an eventual covenant.

If you have a situation where the covenant would produce a reduction in development potential for the grounds because the grounds were quite sizeable, then in the future, the owner may decide to subject the grounds to a covenant and receive whatever tax benefits we hope we have achieved by then.

The legal instrument affecting the designation can be tailored in a highly specific manner to limit the regulatory authority to very specific portions of the property. The case I often site here in Calgary is the designation order on the house of William Aberhart, the former premier of the province. The designation order literally applies to the front facade of the house and in a straight line from the front facade to the property line on each side, and from there to the street. Literally everything from the front facade to the back alley is unregulated. You could knock the whole place down and build a new building behind it if you wanted to without ever having to come back to the Province. However, if you were going to do anything to the front facade and the ground from the front facade to the sidewalk you would need Provincial approval.

The Barron Building

8th Avenue S.W.
Calgary, Alberta

There were three 1910 clapboard Cliff Bungalow style houses. Penley's Dance Academy was directly adjacent to the site, in what still remains as the London Building, which is due west of the Barron building. It is a recently painted three-storey brick building. On the corner, the building that is now the Chicago House, formerly a modern 1964 Bank of Montreal, was Waldren's Used Car lot.

Barron Building, Calgary, Alberta

Discussion of the Barron Building

Presented by Fraser Shaw

What we tried to do was articulate in a very broad way what we thought was significant about the building. We did not deal with the mechanics of the covenant. So we are not aware of all the issues to which the Trust could be exposed. We started out with a character statement. Basically, it was Calgary's first skyscraper. It has an association with the Barron family and the development of the oil industry in Calgary. It was the first office building; with a multi-use precedent, a penthouse and a theatre and retail development on grade. The building has a unique style that is kind of an Art Deco/International Style hybrid, with the application of various concrete slab-pouring techniques.

We identified features of the building's interior, exterior and context that would merit consideration in a covenant. On the exterior we included the façade, articulation, scale of the massing, the Art Deco pavilion and the ribbon windows. Of the ornamentation we included: the Art Deco style, the materials, the Tyndall stone limestone and the terrazzo. The glazing fenestration is quite unique as it wraps around the corners and the marquee over the theatre entrance.

Interior features included the elevator lobbies with terrazzo floors and mahogany panelling, the penthouse with its fireplace and period wood panelling. We also included board-form concrete behind the scenes, and the theatres. Apparently in its day, it was a fairly significant clearspan that represented quite an open space in the theatre lobby entrance.

Preservation planning issues include the set backs and exposed building facades. These features would be vulnerable to any development on adjacent sites. The prominence of the front facade would lose quite a bit if a development were to be put on the corner lot. The east property is particularly sensitive.

The view shed was also examined. The building addresses the south, which was once the south edge of the city, now more defined by the railway corridor, but it still looks that way. That is something that would be appropriate to preserve, in an ideal world.

Long-term code upgrades within the building were also considered. Key to address would be the fact that there is no sprinkler system at this point. The building also doesn't have sufficient parking for city requirements.

Will Plus-15 connectors be required as new development is added to this block? There was also a concern over future treatment of neighbouring buildings. Would they advance toward the street beyond this building and diminish its distinctiveness and significance?

We looked at the scope of the covenant, looking at what would fall on both sides of the scale without really addressing the fine mechanics. Basically it would be a prophylactic against further degradation. We didn't really address the restoration. The owner has looked into the provincial designation program and one of his concerns as the developer/owner is whether it would encourage him to restore to a high standard and whether the available dollars would be sufficient or meaningful for a building that size?

The advantage of the covenanting system would be that it could apply to the features and not the whole building. There could possibly be a tax rebate involved. So the dollars involved would be more favourable for the developer, especially given the long-term development potential of the site. As it turns out, this developer also owns a partnership in the neighbouring property, so there is really good potential for consolidation of this corner lot.

The covenant would be concerned with preserving as much of the exterior form, decoration and glazing on the facade. That would apply to the east, west and south sides of the building. The north could be developed, because this does not face the street and it would give the developer more scope to make better use of the property in the long term.

The theatre lobby, the auditorium, the spaces that are publicly accessible and the marquee would continue to be used as it is today and would stay in its present form. It would not extend to the ancillary spaces, because functional upgrades with theatre companies would require some flexibility in those areas.

The penthouse shell would stay as it is but the feature layout could change. The fireplace, however, is a key period feature that would have to stay. Elevator lobbies would also stay. The terrazzo floors could be periodically replaced, as tenants required their own corporate logos. The intent there would be to add an element of change to the building.

This would be set against a tax rebate on a percentage basis of the flow of the assessment, on an ongoing basis. A lot of the discussion was focused on what would be meaningful for the developer in terms of the principle for him to take on these kinds of encumbrances. Basically, a tax exemption would be the most meaningful thing over an ongoing period, rather than say an initial ten-year period, because of the ongoing maintenance issue.

The owner would provide access to the public areas, the lobbies and theatre. That would be in business hours, on a regular basis. The penthouse itself would be accessible much less frequently, perhaps twice a year.

The Trust would provide an annual architectural review and assessment. At the end of our discussion we started to consider what we would do in terms of restoring the building and to what extent we would actually restore it, because of the various accretions. We never did get that far.

Question: It looks as though you have indicated a preference or requirement for a particular use in the interior. I have found that it is very difficult to legislate uses, because you are restricting the owner unreasonably to what could be viable within the building, providing the uses are not inconsistent with the conservation of the elements that you wish to preserve. Then I would say that you might need to allow some latitude.

Fraser Shaw: What we did was restrict the use of the theatre. A change in use would mean a change in the form. The rest of the building is only the lobby area so it means leaving a blank slate other than the lobby areas.

Question: Are we prepared to countenance the fact that the theatre space might stand empty, and would that be acceptable?

Fraser Shaw: To the owner it is acceptable, because he finds that space is quite precious and he wants to keep it as a theatre. That is why we chose to go that route. I think in the future if a covenant was in place and there was a different owner, we would have to negotiate a different use for it. But, this owner is fine with its being empty. He would prefer to show movies and have the space used but he is in love with this space.

The intention was to make the space viable as a presentation venue. So whatever changes were necessary to make that happen would be allowed. There was nothing so remarkable architecturally about the interior that we would impede that.

Comment: One other comment relates to something we discussed in another group. When you are looking at whether it is exterior or interior coverage and you are choosing some elements and leaving others aside, what tends to happen is that it is a bit like spot welding. If you are not using a holistic approach to it, then whatever you change impacts other parts of the environment or fabric, whether it is interior or exterior fabric. As I indicated that it has been the Foundation's approach in the past to try and take a more holistic approach and probably use the coverage. Make it a global coverage, but provide an indication of where your focus is going to be. Otherwise you get into administrative issues that are often very detailed, and you discover that some things that are happening to the interiors overwhelm what you are trying to protect.

Fraser Shaw: What makes this an interesting situation is that it is essentially a spec office tower and there are several floor plates, and those floor plates could be a single tenant or they could be multiple tenants per floor plate. There are any number of configurations. So we hit upon the idea of conserving the most public aspects and that includes the elevator lobbies. They are a continuation of the penetration into the private portions of the building. So that if, for example, a company wants to establish a strong identity in the lobby spaces, they would be permitted to do that on the wall surfaces, but the terrazzo floor could never be altered to the point of adding a letter or some other element to the floor. I can't think of another use of the inside of the building that could cause that sort of conflict to arise, other than actually blocking out a window, which I think would be covered in the other part of the covenant. In terms of use, we discussed its potential for residential use.

It is necessary to weigh the long-term viability of the building to guarantee that it will be maintained and remain a vital part of the community. I am not sure how best to do that.

I think it is fair to add that the owner told us that there have been a number of renovations. What we really dealt with were the special areas that are left.

On another subject, the owner has a desire to stretch out the tax benefits. One way to do that is to stretch out the donation of the easement. You may protect one facade one year another facade another year, you may protect the interior, another year each time with a new grant of easement and each time that grant is a new contribution to the non-profit organization. Now it is up to the non-profit organization as to whether they think that is appropriate in terms of ensuring the protection of the building. In this case, it is likely that simply preserving the front facade and knowing that the owner is likely to give addition facade easements and say the lobby easements later, may be sufficient reason to take the easement in that form. In that way the benefits could be stretched out over time.

Marc Denhez: I am not familiar with whether that technique has been used in Canada, but I see no reason why it shouldn't be. There are separate carry forward provisions.

Realities of the Development Industry: The Impact of Covenanting on Development

Panel Discussion

Introduction

The following is a panel discussion on the impact of covenanting on development. Presentations by were given by: Mr. Bob Holmes, Mr. Harold Milavsky, and Mr. Don Douglas.

First Speaker: Mr. Bob Holmes, Planning Commissioner, City of Calgary

When one sits down to talk about how the public and private sector can get together and begin a discussion about heritage conservation, the first thing to acknowledge is that we probably have stereotypes of each other. We have to get over these in order to get on with more creative part of the discussion about achieving each other's objectives. I think from my own experience as the Planning Commissioner in the City, and seeing our own heritage program evolve over the last ten years, there has been concern about identifying the buildings of architectural and historical significant. We need a record of what and where they are and who owns them, and establish some initial contact with the owners. I think that is now behind us.

Secondly, there has been concern about whether or not there should be controls. If so, what controls would prevent the premature and unauthorized demolition or alteration of these buildings before the discussion about their future use can be embarked upon? The discussion may occur in the context of a Heritage Conservation Policy or the City of Calgary's Land Use Bylaw.

Thirdly, what are the standards that we are intending to apply in the preservation or rehabilitation or restoration of a project? What are the trade-offs? I believe that the private sector's view is that a heritage building is either an asset or a liability, depending upon its location, its condition and/or the market place in which it operates. Whether or not it can be turned from a liability to an asset depends upon the developer's view of the market place and the opportunities. The developer's concern is about a return on investment. Concern is also shown about the approval and negotiation process with the Planning Authority, and whether there is agreement on the issue of reasonableness and the bottom line, unless the motives are philanthropic.

In my experience, successful projects have reached beyond the stereotype of saying, "When I go to see the Planning Authority. I am going to be taken to the cleaners." On the other hand, the public sector feels the private sector developer needs to be closely watched for the public good. Successful projects occur when the two parties get beyond the stereotypes and activate creative and collaborative problem solving.

I see the covenanting issue as part of a larger tool kit of attacks and standards and market and ground lease and other issues that can help the negotiating parties deal with the issue of achieving preservation or restoration. When the municipal, provincial or federal governments own heritage properties there is of course there is a bit of a dilemma. Everybody is watching to see whether they are going to play by their own rules or whether they are going to conclude that it is all right to regulate the private sector. But when we get our own architects and cost consultants involved at looking at our own buildings, we quickly conclude that it is far too expensive. Restoration is not going to rate as high as building another fire station or putting another ambulance on the road or fixing another bridge or building a piece of highway.

Two of the experiences I wanted to describe involve examples of how the public and private sectors have joined to achieve objectives whose success you can judge. The first project is the Burns building. This is a building in downtown Calgary on a block that was redeveloped in the late 1970s and early 1980s for a new Centre for the Performing Arts. The development climate in Calgary was very buoyant. The economy was booming and a group of very well intentioned and aggressive people decided that Calgary should have a new Centre. They chose a block in the downtown because they believed that the Centre should be the anchor of a theatre district, rather than picking a green field site surrounded by park space and fountains. Of course when they chose a downtown site in the older part of Calgary, it did not take very long for us to get into a discussion of the heritage issues.

There were two buildings on the block that were subject to considerable heritage discussion. One problem was resolved quite easily. There was an old Federal Government Post Office on the site that the architects had chosen rehabilitate and incorporate in their design and use as the lobby for the concert hall. A cynic might suggest that the building was so solid and large that its demolition would have been a major event.

Most of the debate, however, revolved around the Burns Building. It was built in 1916 and is a very fine example of a terra cotta building and is worthy of preservation. So the public debate about the Burns building began. The City owned the building and we were asked to pull some material together to help City Council decide what should happen. We retained Dr. McMordie to give us a report on the architecture and the historical significance of the building, and we retained an architect who was experienced in building renovation to do a feasibility study of the costs of renovating the building.

On a very close vote, City Council decided that the City was not prepared to put any money into the building. The people planning the Centre were not prepared to incorporate it and did not have the budget to do so. Rather than putting City money into the building or demolishing the building, the City issued a request for proposals to the private sector to see if there was any interest in investing in the project.

The proposal call went out and a proposal was selected from Toronto architect Jack Diamond and a group of investors. It proposed that the private enterprise organization would assume responsibility for renovating the building at their expense. That made it easier to deal with the financial side of the issue. The important thing is that the City decided it was not prepared to sell the building. We were concerned that the building should be restored to a high standard. The exterior was particularly important because it was on the block on which our new Centre for the Performing Arts was being built. If we were to sell the building, we would not have the necessary control of the restoration and rehabilitation. That was difficult for the private sector because their security was a ground lease rather than a fee simple interest in the property, but there was good momentum and the project was completed. The ground lease is a mechanism that the public sector can use to assist restoration.

There was also a debate about the value of the ground lease and to what extent it provided an inherent subsidy. If the City had had the ability to use a covenanting mechanism, it may have been possible for the City to sell the building and still achieve control over the quality of the restoration, the preservation of facades and the elements of architectural and historical significance.

The other example is also downtown, but was a reverse set of circumstances. In the early 1970s a Convention Centre had been built, but in the 1990s expansion was needed. It was decided that the expansion should be adjacent to the existing building, so we were again involved in a discussion about integrating new development with original sandstone buildings. Some of these buildings were built before the turn of the 20th century, were provincially designated, and were the finest collection of sandstone buildings remaining in Calgary.

The City owned most of this block and had assembled it during the 1980s and 1990s in anticipation of growing needs in this area. A private company called Balboa had in fact purchased the heritage properties that existed on the block. We found ourselves in a position where the City had substantial ownership of the block, and the only other owner was Balboa. Our building program for the Convention Centre expansion involved a clearly delineated rectangular configuration for the convention hall. There are certain program requirements for this facility that we needed in order to be competitive in the tourist and convention market place.

An efficient design for the convention centre enabled the creation of a surplus parcel of land. We found that the best thing to do was to have a comprehensive design for the development of the whole block. In order to achieve that, we began discussions with Balboa. We agreed to sell Balboa the residual parcel that we did not need for the expansion. We thereby gave Balboa a developable parcel with more than just the heritage building but other vacant property they could use. That led to the comprehensive development of the whole block with Balboa building the Hyatt Hotel primarily on the vacant parcel and proceeding with the restoration and renovation of the heritage building. The City created the Convention Centre expansion.

I think that that project is a case where the negotiating parties were able to get beyond the stereotypes they had of each other and put agreements together. The agreements were to allow the private sector to have a developable parcel of land, while the City accomplished two of its principal objectives: to proceed with the Convention Hall and to encourage Balboa (without providing an enormous public subsidy) to privately finance restoration of the heritage buildings. When the opportunity to build the hotel presented itself it was satisfactory to us because in the tourist and convention market place, people look at the relationship between meeting space and the proximity of hotels.

My point is that we were able to find tools that we could use in the negotiations to achieve our respective objectives. It is important to get beyond the stereotypes of having the public sector with the regulatory sword under the table hauling it out every time it did not get its way, and the private sector saying this is unreasonable. We have enough risk in the market place and have limited tolerance for the additional risk associated with the approval process, without any certainty of a successful product. I think that a covenanting arrangement may have been of assistance to Balboa. I do not know their tax position, but clearly if we had the tool of covenanting it may have come into the negotiations to achieve the development of the whole block.

There are two other things that I think are really important to the achievement of the heritage preservation objective. One of the key things in these negotiations is having good lawyers. I do not mean good lawyers to protect your interests and tell you why you should not do things. I mean good lawyers to protect your interests, and show you how things can be accomplished while doing so. Most people who know about the negotiations on the Convention Centre, Balboa and Hyatt arrangement have spoken about their creativity. Good lawyers negotiated the agreements, whether they were for access to the loading docks for the Convention Centre or for access to the ballroom in the Hyatt Hotel. When you get involved in a very complex development of a downtown parcel of land, it is not possible to say, "You stay on your side of the property line and I will stay on my side and we will just stay out of each others hair." If you do that, you probably miss an opportunity to create a more satisfactory outcome. So you need good lawyers and obviously good architects.

My final point is that you also need to be dealing with people of substance. I don't mean that in terms of character, but in terms of people who have the tolerance to accept some risk, to look at the investment as a long term proposition rather than a quick turn-around, and people who have the sophistication to understand the public objectives and stay in the negotiations for a long time.

We have had our share of tire kickers, who think there is some grand public subsidy available if only they can get into the negotiating food chain. They think that some grand cheque will be written that will take the risk away and make the whole thing look like a private sector deal, but it is underwritten by public money. The successful projects and the more satisfying projects are those where you are negotiating with people who know this is not going to be a quick and easy process. They have an appreciation of the marketability of the restored property, and a business plan of its use. They have assessed the location, and understand that this has the inherent risks of a renovation project. They know that putting the deal together takes longer than building a multi-family project on a green field site in the suburbs.

Second Speaker: Don Douglas, City of Calgary Developer

Covenanting is not well understood by most of us in western Canada, and I certainly welcomed the opportunity to sit and listen to the excellent presentations that were put on yesterday about this whole issue. A bouquet to the Civic Trust for this, because this is certainly something we have needed for some time.

The topic is the Impact of Covenanting on Development. Let me say at the outset that I believe that the impact of covenanting on development will be positive. Over the longer term, covenanting will enable real traction to be made in the preservation of heritage and conservation properties. This is because covenanting, as it has evolved in North America, is designed to enable preservation of both heritage and conservation properties while at the same time offering the land owner or developer an economic benefit for doing so. That is the key.

In the future, the key will be the ease of use and the certainty of benefit offered by the covenanting system. I am a developer, so I am going to take a different tack from Bob. Over the years I have owned many buildings in Calgary that were designated heritage buildings after we owned them. It is important to know where I am coming from.

So why is it that the landowner, developers and the heritage public have historically butted heads over this issue of preservation? The answer is one word and that is "money" – usually the loss of it. Preservation costs are usually borne by the landowner or developer. The imposition of heritage considerations on an owner's land reduces development flexibility and often, though not always, increases development costs. More importantly, unless one purchased the property for preservation reasons, the imposition of a heritage designation could be totally contrary to the owner's original development vision. I think you would agree the act of preservation does not usually allow development to reach the highest and best economic use of the property, hence the cost of preservation and how to determine the value of the easement, because there is a value there.

My job as President and CEO of a publicly-traded real estate company is first and foremost to capture an acceptable return for my shareholders. They are quite adamant about this issue! Sitting through a number of sessions and public hearings, I get the impression that most of you see developers as those fat cats, who should be compelled to carry these costs. Perhaps because of the very nature and scale of the projects you see going up, the question of affordability seems minor. Let me emphasize that the development business is very competitive and the profits at the margin do not allow for unexpected costs or reduction of return targets. When you see these large-scale projects underway, they are balanced very critically at the margin of stop or go in terms of returns. There are many people who are involved in to that discussion – primarily the lenders and shareholders. Preservation costs do matter and greatly affect the viability of a development project. More often than not the developer does not want to destroy all buildings to build some mega project and generate exorbitant profit. Rather they have not yet figured out how to make an acceptable return as well as harbour heritage considerations. That is the issue.

Canada has not had much preservation experience because we do not have the hundreds of years of old buildings, especially in Calgary. Heritage-conscious development is generally more expensive than new development. At a conference I recall talking with a person called Isadore Sharpe, President and Chairman of the Four Seasons Hotel chain, a very successful first class worldwide hotel chain. He was discussing the restoration of a large hotel in Seattle that they had just finished – the old Olympic Hotel. When I asked him about the development experience, he said it was a nightmare and one that he would not repeat again. I think that had to do with a lot of operational issues and trying to make heritage properties work within the competitive world of margins that have to be recaptured.

Yesterday we talked about the designation of heritage properties and the resultant loss of flexibility. As developers we have to deal with these issues as a big cost to us and as significant in terms of what we can accomplish. It was noted yesterday that the designation of heritage properties is rarely done in municipalities. The reason for that is because the municipalities are obligated to mitigate the difference. This tells you the municipality is not prepared to put those dollars into the preservation of heritage buildings because of other concerns and considerations. This is a body that is close to the public and talks on behalf of the public. It is useful to go back to the issue of money and who pays, because designation does have a cost.

It is acceptable if the province pays, because they do not have to mitigate and the obligation is held by the landowner. Historically we have not had a covenanting process that allowed the landowner to recapture any of that cost. If you get notification that your property has been designated heritage, it means you cannot do what you want. It is a major change in your development plans, and that has to be incorporated.

I want to make the point that it is not that developers are insensitive to these issues, but that there has been no mechanism for the developer to share this cost, although there have been some grants. The bulk of the costs are generally borne within the development cost structure. These are significant and can have a material impact on the returns of the project. The costs also have a material impact on they way the property operates. Not all heritage buildings, even though significant from a heritage perspective, are easy to modify from an operating point of view. Problems are the heights of the roofs, the wall, and how you integrate an efficient operating mechanism.

So that is why I believe that covenanting is an important step in this country to enable a merging of developers and heritage groups so that they can come together and find some common ground and go forward with these projects. Covenanting, as it is evolving, does allow for some economic benefit for the landowner.

For covenanting to be a useful tool it is imperative to simplify the process and clarify the benefits. Landowners and developers will be reluctant to use the covenanting process to aid in this capture of lost value if the certainty of the amount of the recapture is in doubt. I would suggest that significant work needs to be done with CCRA to make covenanting work and to ensure that the opportunity that covenanting offers is not lost. It is a fabulous tool. It has been used very successfully in the United States. I have been involved in covenanting a little bit on the other side in terms of conservation properties. It has been very effective, especially south of the border, and is starting to make great strides here in Canada for preserving natural spaces and wildlife areas. The reason it is successful is the tax incentive for the landowner.

If it were not for the tax relief, it would not have much attraction, because you still have the problem of giving up development potential. I am not talking about heritage issues that are desirable in the view of a third party and not the landowner. When you have heritage issues that are also desirable to the landowner, you do not have a problem. There is a commonality of interest in that case. When you do *not* have a commonality of interest, then it is an issue of who is going to pay for the lost value that the covenant imposes on the property. I don't think covenanting goes all the way to recapturing the costs, but certainly it goes a long way to allowing the landowner or developer to say, "This makes some sense. I can see myself doing this." It becomes an affordable issue. I will leave you with the thought that it is not about preservation, it is about who pays.

Third Speaker: Harold Milavsky, City of Calgary Businessman

I come from the development industry where our total focus is office buildings and shopping centres. I can see that covenanting would have some application to the office buildings that were being built downtown which were historical buildings. I do have some experience there over the years. What Don has said is absolutely correct – it is the economics of the project that drives it. Financing from third parties will not be there if you cannot make returns.

The other point I would like to make is that the development of office buildings, which I am more familiar with, is very high risk. Just to give you an example of that, a project like Bankers Hall downtown is a complex that has incorporated a heritage property. The project has taken twenty years from the concept of building Bankers Hall to the time it was finished. The first phase was a tower of thirty-seven stories and eight hundred thousand square feet. It took ten years from concept to building because of changing economic times. The second tower took another ten years to build.

The Hollingsworth building was provincially designated and was built on the corner of 2nd Street and 8th Avenue SW. The Canadian Imperial Bank of Commerce occupies it. It is a terracotta building kept in its historical condition. Inside there was also preservation of a historical elevator and staircase. Those were all built into the project.

We also undertook during that period the restoration of a brewery in Denver, Colorado, without any input from the public sector. It was converted into a retail complex called the Tivoli. Unfortunately it was not a great success. It did not have the full support of the community of Denver and as a result it ended up being a beautifully restored brewery, but the economics did not work out and it was of no value. It is now shut down.

The point I want to make about some of the other tools, besides covenanting, that could help developers to make these restorations a possibility is density transfer from the City. If the developer can get additional density for doing some of these things, it would give the developer more economic return to offset the additional costs of preservation. Coupled with that would be the ability, in some cases, to be able to transfer that density to other projects at a later date. It would have a value, which the developer could use. Don Douglas makes a very good point that a developer has to feel some certainty that they can use these covenants before they are tested by CCRA. If you take the risk and do a restoration project because you think you have tax relief, and then CCRA decides you don't, you have a real problem.

Another issue with regards to covenanting is its implication on municipal taxes. They should be less. The difficulty with this, is that if there is a lesser value, there is no consistency across the country in how you measure those values. Calgary is considerably different from Edmonton. Calgary uses a method, I believe, where they assess a building as a double A or B or B+ and regardless of the economics of that particular project. Edmonton takes into account the revenue stream that comes from the project.

From the developer's perspective, most office buildings are net leases. That means that the proportionate share of taxes for a tenant is passed on to them. So if you do get a benefit of a reduced value on the building and therefore lower taxes, it ends up going back to the tenant and none of that is left with the developer. That is the case if the tenant is in existence at the time. It may have an effect on the ability to get an extra dollar of rent from the tenant. These are the kinds of things that would have to be worked through.

Discussion

Question: Economics drives any project, and the question in my mind is that if there is a value there, the public sees it but will not pay for it. The issue is to have the public pay for that margin. I would be open to any suggestions as to how to persuade the public, which means municipalities, provincial governments, federal governments and so on. Mr. Douglas said it has to be clear and consistent. Is it then our preferred way to persuade the federal government to provide an opportunity or is there any benefit to developers to have municipalities pay?

Don Douglas: There could be many ways to enhance sharing the value of the cost of preservation. If we say preservation is important, and it is important to the public rather than a particular landowner, there is always the capacity for municipalities to do that through expropriation. Most developers will tell you that they are prepared to go through that. There is a process to get you back to fair value, if you do not want to go along with the concept. We would certainly be open to new ideas on how we could share that cost. We own a number of buildings along 8th Avenue, over many years, and have been trying to make a dollar out of those things. Then someone says it is now a heritage building and it totally changes. We have no argument (while wearing my heritage cap), but the question really becomes how we deal with the loss of opportunity. As noted earlier, if you get a 7 FAR density on a site like that, it has value as density transfer. In Calgary, however, where the office environment has not been robust lately, density transfers are not something that anyone would pay for. We have pretty high-density opportunity anyway. It has not had a dollar value except in circumstances where a site might have been constrained, or there is a technical issue; then it becomes valuable.

There could be many ways the province and the municipalities could find a way to share the cost. I think they should do that. One way is property taxes but there are other ways, such as giving a tax holiday for heritage buildings for a certain period of time. The real issue is simply one of money and value.

Question: Would you care to talk about income tax measures?

Don Douglas: I think income tax measures for many people are not valuable if they are not taxable, i.e. the Roman Catholic church is non-taxable. So here we have somebody who can keep an asset from someone who could develop it and we have the heritage folks saying, "That should not be allowed to happen." Covenanting likely might not work there, unless you can transfer the value of the covenant to a third party, which again would be very valuable. If in fact you could covenant that site and then sell it to someone who needs the tax incentive, I think that would go a long way towards what the Diocese would need in terms of revenue stream, because I am sure that is the issue. These are really municipal issues.

Bob Holmes: I will briefly comment on the property tax situation from the municipal point of view. Unlike the federal government, in Calgary and other municipalities in Alberta there has been a reluctance to use the property tax as a mechanism to achieve other objectives. We have made a painful conversion from an old and antiquated way of establishing property tax value, through discounts and things like that, to a market value form of assessment. It has not been easy, but most people say, at least the principle is clear that it is the market value of the property. But if you start putting on the back of property tax system an exception for this or that, then you really are loading up the tax system with a bunch of other objectives that it was not designed to accomplish.

We do have mechanisms for providing tax exemptions to churches and other non-profit organizations. The municipalities' ability to do that is defined by the *Municipal Government Act* of the province and the regulations that flow from that act. I think you will find most municipalities acknowledging when the tax would apply and whether the tax could be applied on the basis of the percentage of occupancy of the building. To say that you have invested in heritage property, therefore you are going to get a ten-year tax holiday, on the face of it, it sounds like a good idea for this objective, but then you have people coming in and saying, "What if we have noble objectives? Would you give us a tax break if we did this?" That is when I talk about overburdening the system and complicating it. Look at the complications that are in the income tax system such as the loopholes, and people's concern about wanting some certainty.

Question: What is the complication with municipal taxation? What are the reasons for which the property tax is modified?

Bob Holmes: The point I was making about tax exemptions is that there has been a long established acknowledgement that for certain types of organizations, either defined in the Act, or defined in the regulations arising from the Act, or with some discretion from the local municipal council to declare you tax exempt, or we choose to treat you as tax exempt, if you are a church or non-profit organization, you fall within that definition.

Harold Milavsky: There have been pressures on cities to give tax breaks to get this particular industry. That is a mug's game, because everybody looses.

Bob Holmes: That is a good example because you can get into the business of providing tax exemptions for high technology companies to settle in Calgary, or for this, that or the other thing. I am not dismissing the idea, but I am simply commenting on the complications of using them. I do acknowledge the fact that the property tax system should be reviewed from time to time and if there are municipalities in Canada that do use the property tax system as a way of dealing with the cost issue (the cost to the owner) assuming the owner is not the City, then that helps.

Don Douglas: There are other ways we could look at that. How important preservation is to the public? Will the public put up their dollars through the public system? In cities like Denver for example, they have a surcharge for the arts centre. They did this through plebiscite. We would definitely like to have something like this here. They have a one percent annual surcharge. The money is used to sustain the arts in the community. Heritage could be part of this. The province is fortunate in having a good cash flow and will be debt free in about two years. We could establish foundations, and fund them for projects that are for public good. Heritage would be one of them. Perhaps you could establish a heritage foundation that was funded by both province and municipalities either through a surcharge from the province and through the municipalities with these funds being administered to foster preservation. I find in interesting in England that they don't use the tax system – they use a public collection system and it has been very successful.

Audience comment: At the risk of simplifying the matter, I have been very interested in listening to this particularly as I am from eastern Canada, where money is rarely an option. I am particularly interested in the mechanics of density transfer. It seems to me that there is no argument that people are going to require some compensation for the loss of certain rights over their property. Financial compensation through the tax systems seems to be the option of choice, but there is the discussion of the exchange of rights. I am from Halifax where density is an option and we have this non-renewable resource of dwindling heritage buildings. What we also have is a very constrained geographical situation – the heart of the city is on a peninsula. There is a lot of discussion about increasing density on the peninsula, particularly because of transit needs. The mechanics of density transfer interest me. It seems to me you are talking about discussing density transfers which can be for a current project or for a project at a later date. Is there a futures market in density transfers? How does that work?

Don Douglas: The experience we had was that we did have some density given to us by the City in exchange for putting in a Plus-15 bridge where it was not required, and then we used the density ourselves and put it on a development that we were doing, Western Canadian Place. But there is an agreement between the City and the developer and you can make provisions within that agreement that it can be sold to someone else and they can put it on their project. It would be under the terms of the agreement that you had between the City and the developer.

Audience Member: Is there a limited term of years in which that could be saleable?

Don Douglas: No, there was not. If you did have limit it would have to be fairly long. As I mentioned before some of these projects take a long time to mature. If time were short, there would be no value.

Audience Member: I assume there are other sorts of rights as well as density requirements. I am not sure just what they are. Do you have any examples of other kinds of rights that have been given up in exchange for heritage protection?

Don Douglas: No I haven't. But other things have been done in the United States. Trizec at one time owned 25 per cent of a company known for its redevelopment of downtown areas. Generally these were former market streets that were redundant and were being restored. They got a lot of subsidies from the cities, or in some cases from major business, who had an interest in seeing their downtown restored. For example, in one of the cities a major insurance company as part of their "give back" to the city helped subsidize a particular project. That helped it succeed.

Audience Member: Is there a classic financial model?

Bill Byrne: In looking at this whole issue of taxation and creative ways in which it can be used to provide incentives for various things, I cannot resist by beginning with a news story that I saw on the weekend, about a small town in Saskatchewan, east of Saskatoon, which announced a property tax incentive program of $200 for each child born to households in the community over the next ten years, in an attempt to bolster the sagging population of the community. Bob, I think that flies in the face of everything you have ever looked at!

In any event there is obviously a tremendous amount of interest in the discussions that we have had in the last couple of days relative to this. I think it is fair to say that basically we are preaching to the converted in terms of the value of these things, and there are long-term potentials. I think the issue comes down to the fact that for all of this to work, it has to be reasonably transparent, reasonably understandable, and certain. I think the certainty is the most critical component. It is going to be the most difficult issue. Not only the certainty in terms of making it part of the standard system, but that these types of covenants will have an assessed value that has some meaningful values associated with them so they can influence decisions on a project viability. That is going to be tricky.

We all know there are a lot of different models for projecting the potential values, but it is going to take a long time to both deal appropriately with the taxation departments and policy on federal, municipal and provincial levels to see where these concepts are actually accepted. The evaluation process itself is also going to be trickier than some of us anticipate.

The general presumption is that with some kind of heritage covenant or heritage designation, there is a loss. Part of the complicating factor in this is that it is not always going to be the case. In some instances there are going to be situations where designation and protective covenants and easements actually enhance the value of the properties. I am reminded of the only test case that I am aware of – I think it was the University of Western Ontario looked at about 110 properties that have been designated in Ontario over a period of time, then compared them to the property values in the same neighbourhood, same property or as close as they could come, and the interesting thing that was found was that, over the course of ten years, the property values of the designated properties appreciated by approximately 17 per cent more than the non-designated properties. I don't think that is the case across the board, and certainly within the study that they did there were lots of properties that lost value. It was just that other properties increased in value.

One of the things that interests me is the issue of municipal designation, particularly where there has been great reluctance to proceed with any municipal designation because of the potential compensation. But the truth of the matter is that nobody has ever tested it. Nobody has run a hypothetical model through an evaluation process and through an evaluation board that would set what that compensation might be. There is no question that the provincial statute that allows for municipal designation has serious flaws in it. The compensation clause states that the compensation must be in cash. You could not compensate municipal designation through tax relief unless the owner agrees.

Marc Denhez: The really nice thing about a workshop like this is that it gives various stakeholders the opportunity to dream in colour for a while not about the situation right now, but what the wish list is for the future. To follow up on Bill's point, if we can think about where we are now compared to where we want be in the future, the message that comes out crystal clear from your presentations is that in order for heritage to be sustainable over the long term, like every other industry, it has to be cognizant of the laws of supply and demand. What you are telling us is that the cost of supply is still quite high, or at least perceived to be quite high in the Calgary context. The demand is still quite low in the Calgary context, and this is not going to be a sustainable industry till the demand curve meets the supply curve and eventually exceeds it. Until that point there is a gap that has to be bridged and so the discussion in the last few minutes has been on how we bridge that gap. Do we bridge it through the transfer of development rights? Do we bridge it through grants? Do we bridge it through municipal or federal tax measures? Of course, all of this we would like to think is in the short term or interim period. In the long term, we would like to think that the demand curve is going to be sufficiently strong so that it meets and eventually surpasses the cost curve. Now, there are other cities that have reached this point. Montreal is there. St. John's is there. In St. John's, the head of Tatum Oil made arrangements (the demand for heritage space was so strong) that he parked himself in the first million-dollar condo in a heritage building in the city. In some places the demand and cost curves are where you want them to be, but they are not there in Calgary yet.

I have three questions. We know that there are cases where you can reduce your cost dramatically if you hit on the right the people who have the right expertise to bring heritage projects in on a better budget than other people. What is your take on the availability of that expertise in the Calgary market? That is question number one.

Question number two. This I find a little puzzling. When a buddy of mine did a heritage building in Ottawa and he wanted to market the great exposed surfaces of brick wall on the inside and wanted to market the wood beams and all those other various things, and he wanted to find the best expertise in the country to be able to sell heritage – he actually hired a firm from Calgary to do that.

In another case, the people who did the Sinclair Centre in Vancouver then went to work for a firm that trains facilities managers in how to really be attracted to a lease in old buildings, preferably heritage buildings that have been adaptively reused. Now where is the company that trains facilities managers in how to lease old buildings? Well, they are also based in Calgary. Yet, I don't know of anybody in Calgary who has ever heard of them. So my second question is what is being done in order to raise the demand for heritage properties in Calgary, to raise the demand for class A space to be installed in heritage buildings?. My third questions is; What, from a business perspective, do you see as the opportunities to bring those various forces together so that eventually, maybe not in a year from now, but hopefully in five or ten years from now, you may be able to bring the market to a point where the demand is sufficiently strong and the cost expertise is sufficiently available that you will – without crutches, without various props and perks and quirks – be able to bring heritage properties on the market at a price that the market is prepared to pay for?

Bob Holmes: I think we will hire Marc for the Economic Development Authority in Calgary. Marc, you have just punctured the stereotype that we understand easterners have of Calgarians, and I always like to hear that because it gives me the opportunity to rub in the face of some people that I used to know (since I grew up in Ontario, born in the west) that we also happen to be the city that has the highest percentage of people who are taught in

A.E. Cross Garden Café, Calgary, Alberta

French. Thanks for educating the audience on the degree of sophistication in Calgary.

I do not know if there is a separate market for heritage properties or not. I said earlier on that I think that the location has to be part of the consideration. If I wanted to get into the restaurant business, I don't know whether I would want to get a heritage property wherever it is, as long as it is a heritage property, and renovate it and hope the customers will come. I have got to look at the location and a variety of other factors besides that. My own observation is that there is a growing appreciation among Calgarians for the value of that kind of interior space. But still the discipline is the discipline of the market place. We own a heritage property in the community of Inglewood (A.E. Cross Garden Café) and it is operated by a very fine restaurateur. He has a lease with us and he can just make his lease payments. It has nothing to do with the fact that he is in a heritage building. That is an asset. It has nothing to do with the fact that he is not a good restaurateur – he is an excellent restaurateur. But he is not in a good location. He is surviving on a seasonal basis.

Harold Milavsky: I would say a couple of things, Marc. One, I think what has been done on 8th Avenue in recent times is very good work and I do think we have very skilled workmen here and they are as busy as could be given other things that are going on in this city right now. So, I think the expertise is here for restoration, if the economics are here to do it.

The other thing that is an issue is that we are still so young. We will be having our 100th anniversary in 2005. We were tearing down twenty-five year old buildings to build new ones. Much depends on who owns the property and why they bought it. I haven't seen (other than on 8th Avenue, 10th Street in Kensington and 17th Avenue) where streets have been maintained the way they were.

Bob Holmes: We have a lot of expertise as it relates to restoration projects and how to sell them. The issue that Harold mentioned is a key one as it relates to heritage. We don't have lot of heritage opportunity in a city like Calgary, compared with a city like Winnipeg. I just am awestruck each time I see the number of old warehouses there. What an opportunity. You can buy them amazingly cheaply. I think the issue is not the building – it is the area. We do have a few small areas in Calgary, which were largely old buildings. You can look at from a building-by-building basis, or owner-by-owner basis. You have to designate the buildings as an area we want to keep its personality. That takes front-end money and leadership at the municipal level. If you can get the momentum going in these areas, as demonstrated by Kensington in Calgary, you then create a demand to be in the area.

The problem we had for many years on 8th Avenue was that nobody wanted to be there. Even if you did one building on the street you still were one small person in a sea where no one wanted to spend time. So you need scale, and you need momentum. When you get those things you also get the demand to repeat this. So we have to find a mechanism to create this kind of momentum in scale at the front end. The market will take care of it after that. You see it all the time with housing in certain areas of the city. The heritage houses in areas like Mount Royal sell at a premium, not at a discount. In east Calgary, it is something the city is wrestling with right now. There is some product worth keeping and the question is really what the style of this area will be. What is the area concept? If you create that and it has a heritage theme or package to it, I can tell you it will be very successful once it gains momentum. But somebody has to take the risk at the front end. Frankly, most

Pilkington Warehouse, Calgary, Alberta

heritage properties are not owned by the Trizec Corporation. The big developers are not in many of these smaller areas off the beaten track. Inglewood is another area in Calgary. You need somebody who can go in there and amass and create the front end.

Marc Denhez: So you are saying a way has to be found to get the development community to move in packs.

Don Douglas: Yes, at least to put the front-end infrastructure into place. The large problem we had on 8th Avenue was that for years and years you could not rent them because, quite frankly, nobody (including the public) wanted to be there. The moment you turn it around to where the public wants to be there, the opportunities are endless. You have to make it attractive.

Rob Graham: I would like to correct an implication that Marc made when he said that he did not feel Calgary was at the same place as Montreal is in terms of creating destinations or desire for the retention of heritage property. What is happening in the warehouse district in Calgary is entirely driven by the market. The City has had nothing whatsoever to do with it. There have been no grants or incentives. Most recently, the Pilkington

Warehouse just at the north end of Victoria Park, which has been derelict for more than a decade, has been fully restored by SMED and is about to operate as a dot.com e-commerce operation. They then looked around at the next two warehouses. There are only two left, given that the market has already snapped them up and turned a dollar on them. SMED is prepared to go into those as well. So there are developers who are looking for older buildings in Calgary and the province. They want buildings from the turn of the century, our heritage buildings, to make a statement about the business they are in.

A&B Sound Building (former Bank of Montreal building), Calgary, Alberta

Don referred to Stephen Avenue and the years that United Management held property to no avail there, but ultimately is was a company out of Vancouver, A&B Sound, who looked at the traffic on Stephen Avenue and First Street west and said, "Wait a minute, why is this derelict?" This is the best location in the City for what we want to do. Six months later their receipts were as high as they were in their Seymour store in Vancouver. Similarly, Inglewood has turned around in seven years. That is the market, although the city and the province did give a gentle push in terms of grants and putting a program on the street. Once the first baseline projects were in place, the rest of it has been driven by the market.

So I would disagree with Marc. I think there is definitely a market here. While there are isolated cases, and there always will be, such as St. Patrick's or the Lougheed building where there are problems where the developer purchased it for a specific reason and doesn't wish to be thwarted. That said, on balance, Calgary is not very different from the City of Montreal.

Nora Robertson: I am speaking as a heritage preservationist. In terms of development, I certainly have sympathies and I think that the tools that have been discussed these last couple of days are excellent.

Calgary is a very young city, but having said that, our heritage buildings are a non-renewable resource. They are our past. They represents what the vision of the city was when it was built. We have to keep that in perspective as we look to a new vision for this city. As Rob was mentioning, there is a resurgence and desire for people to have office space, home space, and restaurant space in old buildings. Why is that? Because it has character. I believe that the downtown core was destroyed in the 1980s with the new office towers. They overshadowed the walking streets. When I was a young girl the 8th Avenue Mall was a very vibrant walking space. I always went downtown to shop. I went to the Bay and Eaton's and as a young child it felt very comfortable to be there. As time went on, and as the office towers were being built, a lot of the shops were destroyed and the walking space was destroyed. It was dark after 5:00. We still have the problem that the downtown core empties out.

With the resurgence of the Stephen Avenue Mall being redeveloped, where restauranteurs and shops are wanting to be part of that, it is being revitalized. People love the character of heritage buildings. People love to go to Europe. People love to go to eastern Canada, where there are heritage buildings. If we destroy every single one that we have, or what little we have left, there really is no reason to go downtown or into Inglewood. Inglewood, thank God, was left derelict for years, because no one wanted to go there, so it survived. Now people love it. People are dying to rent and buy property down there. There is huge value in heritage property and once we see that and developers especially begin to appreciate what value there is there, I think you will see the rewards reaped in the future. Just hang in there. It will happen. We just need these tools. Whether it be covenanting, tax breaks, I know our Society and Heritage Canada have been asking the Federal government for tax incentives at that level. We all need to look at it; we can't keep passing the buck, from municipal, to provincial to federal. We all have to start working together, and I think this is an excellent opportunity to start doing that.

Audience Comment: I have a comment and a question. A comment for Mr. Holmes, to say that I think that heritage is different when it comes to tax incentives, because it is non-renewable and because it is a part of what makes as a city, whatever city that is. I think it does create a different rationale for giving some kind of tax holiday or tax break.

The question is for Don Douglas. You spoke very briefly about nature conservation. There was some comment about the federal tax system having some form of tax relief for people who make donations in the natural conservation area. How has that group managed that? What do we need to learn from them? What kind of process do we need to go through to have a similar kind of break for heritage?

Don Douglas: There are probably some folks here who could give you a better answer than me on the process of getting that in place. I can tell you that the tax benefits that have been available now for conservation properties, wetland properties and nature preserves have impacted many properties here in Alberta. Just recently, the Nature Conservancy took over the Palmer Ranch in Waterton National Park. Again, it is a tax driven incentive for a covenanting situation. It has really been led by the United States for a number of years and now it has come up here.

Harold Milavsky: I agree with the speaker who felt developers destroyed 8th Avenue. Unfortunately, at the time some of those developments were taking place, there was no bylaw and it is one the city should consider for 8th Avenue. Other cities have it. Eighty percent of the street should have to look outward as opposed to looking inward. Projects like Scotia Centre (which I was involved with) and Bankers Hall were turned inward. There were not too many stores along there with an entrance on 8th Avenue Mall. Also in assembling the land (some were owned by Financial institutions previously) from financial institutions you had to agree to put them back in the property. If the City had heritage laws, that would prevent that, although there would be compensation issues.

The last point I would like to make is that in a number of years from now, hopefully Bankers Hall will be a heritage building. It took twenty years to build, so I think it will be a heritage building of which we can all be proud.

Bob Holmes: It is a lot easier to implement public policy when there is a market for what you want to achieve. It is a lot more difficult to implement public policy when there is a perceived imposition of a cost and uncertainty or risk without a market. That goes back to what I said earlier: that those who are negotiating need to take the time to understand each other's point of view and see if they can find common ground to put a project together. That is where the ideas (whether it is ground leases versus sale, whether it is covenanting, whether it is density transfers, whether it is tax arrangements), are part of a growing tool kit that is available to achieve some of these objectives. I do believe, and I say this more as a Calgarian who has lived here since 1967 than in my official capacity, that there is a growing market for heritage properties. I think we have some outstanding examples both in the public and the private sector: buildings that have been renovated and are desirable both from an architectural, preservation point of view and from a business point of view. A lot of factors go into achieving that and sometimes the cycles of public sector interests and private sector market opportunities don't coincide. When they do, it is a little bit easier than when they don't.

Stewardship through Covenanting: Moving Forward

Concluding Panel

Presenters were individually asked to make concluding observations or reflections on covenanting and heritage preservation.

Dr. Bill Byrne: I think we would all agree that it has been an extremely productive time. I would like to briefly make a couple of points.

Let me just say unequivocally that I think that the pursuit of covenanting as a tool to assist in heritage preservation is an absolutely critical component of the movement these days. I think we have been disabused of any notion that this is going to be a simple process, or disabused of any notion that the process is one in which we are going to enter into a partnership with the tax barons of the country, but I think the prize is worth the game.

I say that from the perspective of somebody who has been responsible for administering heritage preservation programs for the better part of twenty-five years in the Province of Alberta and following the traditional approaches of using the various funding schemes. I have long since come to the conclusion that while funding schemes are a part of the arsenal they will never be the complete package. One issue in particular that I noted this morning that Don Douglas raised when he spoke of the possibility of developing some kind of private/public partnership program in a foundation that would develop a funding pool that could be used to offset cost of developments. I spoke to him briefly afterwards and told him frankly my personal conviction is that it will never happen for a number of reasons. By and large, this country does not have a history of private patronage that is similar to what exists in the United Kingdom and the United States. This is a different society and our solutions will have to be solutions made in Canada. I just do not believe that we will ever develop a philosophy of private philanthropy that will generate the amounts of money that would be necessary to have a significant impact on heritage preservation, particularly as it applies to these large-scale commercial projects. How do you deal with the large-scale commercial programs in a downtown Edmonton or a downtown Calgary?

I do not believe that we will ever develop funding schemes in this country that are appropriate. Again I look at the public funding in this province through the Alberta Historical Resources Foundation, in which the government provides all the money to that organization. The reality is the monies are never remotely sufficient for the task at hand. The finite amounts of money, in terms of the global pot, that the Foundation can draw on have not essentially changed for over ten years. The same amount of money that was available in 1990 is what is available in 2000. And yet, the numbers of properties that theoretically would be eligible to make use of those monies, as well as the costs of dealing with properties, have escalated dramatically in that time.

The other thing that I do not particularly like about funding schemes of this sort is that they are ultimately discretionary. That means when you enter into a project, you can't work out the economics ahead of time with any degree of certainty. Again, we have heard from the developers repeatedly that certainty is one of the things that they need, because they have to factor in all the elements up front. They simply cannot afford to take a gamble on a development. They have to have certainty. Covenants have the benefit of that certainty. If we can make the arguments successfully to have the tax regimes recognize the commercial value of covenants and to provide tax relief accordingly, it gets rid of that element of discretion and it turns it into a more formula-based arrangement in which all of the control then lies in the hands of the property owner and the potential developer. This makes life infinitely easier for all concerned and provides the degree of certainty that makes it effective.

I do not pretend that covenants are the sole answer, but designation is not the sole answer and funding through foundations is not the sole answer. There are no single answers to all of this. Rather we need a suite of tools that can be applied to the specifics of a particular circumstance. Speaking from the Province of Alberta's perspective I think covenants are an area we have not explored vigorously enough, so I welcome this initiative of the Calgary Civic Trust, who have been leading the charge in this province for the last couple of years. I think, overall, this is potentially an extremely beneficial tool that can have real and lasting benefits for the heritage preservation movement.

Marc Denhez: I have had the pleasure for twenty years of working with the two gentlemen to my right and I must admit I was struck by a comment that I heard from the gentleman on my far right when he observed to me that to the best of his knowledge, there were forty thousand pre-World War II houses left in the Province of Alberta. It was his wish that by the time his grandchildren could join the heritage movement, there would be some of those left to witness. I was struck by that, because it pointed out something which I think is fundamental to the nature of what we do and that is that the designated heritage properties in the Province of Alberta were never intended to be, are not now and never will be, the totality of all the older properties that we have a legitimate interest in. They only represent a minority, but they represent obviously the elite of the minority. It is a minority of a number of buildings, which in fact, is very much larger. The problem that has existed in Alberta and throughout the country is that virtually all of these building were perceived for decades to be under a death sentence. The nature of our movement has been to try to find one way or another to stay that sentence of execution, and to bring those buildings back into what we knew to be a tremendously valuable potential and tremendously valuable role that that immense investment could play again in the future.

As we tried to assemble that suite of tools that would be able to do that, we would cobble together whatever we could find. We would have the most extraordinary Rube Goldberg techniques to try to save heritage buildings and to re-integrate them into the economy. Many of those Rube Goldberg techniques are with us to this day. Some of them look a little strange but they are necessary components for now. Let us remember what our ultimate objective is to make the reuse of these valuable older components of our building stock as natural and integral to the life of this province as breathing. We have to make the reuse of older buildings part of the main stream, rather than the exception and make the tools part of the main stream rather than the exception. Covenants are exactly that. Covenants are a tool that has been used as easements in various other parts of society for a very long time. So all that we are doing in this step is taking established tools and applying those tools to our own needs. When we look at the tax treatment of covenants and easements, what we are talking about is not the introduction of something new and aberrational. The traditional treatment of covenants and easements is exactly what we wanted and exactly what we could have used.

So what we are doing here is not trying to depart from the main stream and move into exceptions, but the other way around, to depart from exceptions and move back into the main stream.

I think that what we are trying to do for heritage is to make heritage into a sustainable industry in the Province of Alberta as elsewhere in the country. That is the overall strategic vision that this Province should be adopting.

Tim Butler: I have been very impressed by the enthusiasm and by the ingenuity, which I have seen over the last day and a half. If I have one thought it is probably this: that there is clearly substantial potential for using the covenant as a tool to achieve the conservation objective. But there is a hearts and mind battle to be won in various constituencies, government, the public and the development community, and that it is important that the development and concept of the covenant move forward hand-in-hand with winning the hearts and minds battle. I think what came through from the panel discussion first thing this morning was that there was a way to go yet, in terms of persuading the wider community that heritage is important. Until progress is made with that, no matter how good the tool, it will still be a struggle.

Jeremy Collins: The first comment I would like to make is that no one said that heritage covenants were going to be easy to market. It is quite clear that even when there are willing minds and willing parties who want to enter into them, there are still many issues to overcome, in particular the strategies and tactics that are required. The technical nature of these makes them difficult to approach, but the enthusiasm shown in the room indicates that there is a lot of potential out there to start protecting heritage properties with easements. In the Foundation's early years when it was taking easements because of grant conditions, it was an easy approach to building valuable experience and a varied portfolio. I think we have seen quite an overview: that a number of situations exist; how easements can work within that variety of situations; how they can work in conjunction with other forms of preservation such as designation, where there is a willingness to use, in creative and ingenuous ways, each tool to its full potential.

Paul Edmondson: I talked a bit yesterday about some of the administrative issues and problems that we have encountered. I do not want you to get bogged down in the administrative side of this and the fact that there are institutional costs. It is an incredibly useful tool for preservation and it has been effectively used in the United States and it can be effectively used in Canada.

Obviously the stumbling block is the tax issue. That is the fuel that drives the program in the States. That fuel is missing here, or is has been missing and hopefully it will change. I think there is an incredible inconsistency of concept that designation requires compensation because the owner is giving up value of his property, yet an easement or covenant is considered by the tax authorities to be valueless. I think that kind of analysis is not going stand very long. I think that the panel this morning really shows the way here.

The heritage community in Canada is, I suspect, about as powerful a lobby as the heritage community in the United States, which is not saying very much. But the real power in politics is money, as with so many other things, and I think that opening the eyes of the development community, getting the development community people here to focus on the success stories in the United States is important; where easements and covenants have made a significant difference in being able to get that margin cost up to where it is a go, as opposed to a no go.

The reason the Ritz Carlton is doing a rehabilitation of a historic power plant site in Georgetown, Washington, is because of a number of things, but one of the tools in the toolkit is easements, which give tax deductions. So I think linking the development community to this effort is essential and that perhaps, with their support, the ideological arguments of the tax authorities will be quashed.

Biographies: Speakers and Panel Participants

Speakers

Tim Butler

In January 1999 Mr. Butler joined the National Trust as a Director and head of its nineteen-strong legal department. The great majority of the work for the department is property related. Given the nature of the land, which the National Trust acquires, the titles to the land tend to be significantly more complex than on most UK land transactions. His roles are split between leading the department, sitting on the National Trust's management board (the senior staff, as opposed to trustee, body), advising on the National Trust's constitution, and dealing with general legal queries.

Mr. Butler started working with Lovell White & King (now Lovells), a firm in the City of London, in 1982. He qualified as a solicitor in 1984, working in the commercial property department. In 1987 he moved to Townsends, a medium sized provincial practice, starting as an assistant solicitor and ending up as head of the firm's Business Services Division. At Townsends he dealt with a broad range of property work, particularly landlord and tenant and development matters. He also co-founded the firm's construction law unit and advised on the application of VAT (a UK sales tax) on property transactions.

Both at the National Trust and at Townsends Mr. Butler has lectured and presented, to fellow professionals and to members of the public on a range of land-law related topics.

Jeremy Collins

Mr. Collins is Coordinator of Acquisitions and Dispositions for Ontario's Heritage Foundation. Since 1994 he has coordinated the Foundation's acquisition and disposition activities regarding both fee simple and conservation easement property interests.

Mr. Collins received a BA (Honours) in history from the University of Toronto and an LLB from Queen's University in Kingston, Ontario. After several years practising law, he began working a the Ontario Heritage Foundation in 1988 as the Administrator of its Easements Program, negotiating the Foundation's acquisition of new conservation easements. The Easements Program is a multi-faceted program focusing on the acquisition and management of conservation easements to protect built, natural and archaeological heritage resources.

During his years at the Foundation, Mr. Collins has participated in policy development and provincial legislative initiatives promoting the expanded use of conservation easements in the conservation community. He has also been involved in stewardship activities related to conservation easements.

Marc Denhez

Mr. Denhez is a consultant, lawyer, author and public speaker on cultural heritage. He has had some three hundred works published in seven countries, and various works have been translated into ten languages. Books include: The Heritage Strategy Planning Handbook, Heritage Fights Back, and Capitalizing on Heritage, Arts and Culture.

Mr. Denhez has given planning courses at three Canadian universities and at the Academia Istropolitana in Europe. He has developed downtown Revitalization plans, legislation and contracts at all governmental levels, and program evaluations. On the subject of his book The Canadian Homes, the late Charles Lynch wrote: "If you live in a house or own one or build one – if you have a roof over your head – read this book."

Marc Denhez has lectured at almost every university in Canada, at various foreign institutions (including the Smithsonian), and served on assignments in thirteen countries. He has received an award from the International Association of Business Communicators, and a National Heritage Award from the Government of Canada.

Paul W. Edmondson

As Vice President and General Counsel of the National Trust for Historic Preservation, Paul Edmondson oversees all legal services for the organization, including in-house corporate legal services in support of the broad range of programs and activities carried out by the National Trust, its regional offices and historic sites. He also directs the National Trust Legal Defence Fund, the litigation and advocacy program in support of the historic preservation. He has worked extensively on legal issues pertaining to the protection of historic resources in the United States, with particular attention to constitutional issues. He oversees all legal aspects of the National Trust's real estate and easement programs, including issues and transactions relating to the protection of land through acquisition and administration of both fee and less-than-fee interests.

Mr. Edmondson is Managing Editor of the National Trust's publication on legal developments, the Preservation Law Reporter, and is also designated as Corporate Secretary of the National Trust, and is thus responsible for overseeing its bylaws, minutes, and other corporate governance matters. He has served on the staff of the National Trust since 1987. Prior to joining the National Trust, he was a senior attorney for the Federal government. Mr. Edmondson received his undergraduate degree from Cornell University in 1976, and his law degree from The American University in 1981.

Doug Franklin

Since 1983, Mr. Doug Franklin has held the position of Director of Government and Public Relations with the Heritage Canada Foundation. He is responsible for the Foundation's programs in advocacy, public policy and media relations. Mr. Franklin has worked on several significant initiatives of the Foundation, including the federal *Heritage Railway Stations Protection Act*, passed by Parliament in 1988. He took the lead on legislative strategy, and he appeared before both the House of Commons and the Senate Committees as the principal witness in support of the Act. In 1990, he was seconded to the federal government to conduct a study on Heritage

Trusts. Mr. Franklin frequently serves as a speaker and seminar leader for heritage organizations.

Mr. Franklin received his Masters of Arts degree at the University of Victoria where he studied under the noted architectural historian, Dr. Alan Gowans. He became a private consultant in building conservation and preservation in British Columbia. From 1979 until 1983 he was the founding academic administrator of the Cultural Resource Management Program at the University of Victoria. Mr. Franklin was appointed to the Saanich Heritage Advisory Committee in 1977, and served at its Chairman in 1979. He was a Trustee of the CFB Esquimalt Naval Museum from 1979 to 1983, and was the first resident curator of the Capital Dodd Historic House (1859) in Saanich.

Mr. Franklin has been an active member of the Society for the Study of Architecture in Canada and has served on the advisory committee of the Centre for Canadian Heritage Trades and Technology, Algonquin College of Applied Arts and Technology, Perth, Ontario, since 1999. He also serves on the Advisory Council, Central Experimental Farm, Ottawa (a National Historic site).

Mr. Franklin has written and collaborated on several books. He wrote Early School Architecture in British Columbia; Victoria: A History in Architecture with Martin Segger; and Exploring Victoria's Architecture, also with Martin Segger. For his research and writing on the history of coinage design, he received the Guy Potter Literary Award from the Canadian Numismatic Association in 1984.

Robert Graham

Mr. Graham is the Heritage Planner for the City of Calgary, a position he has held since 1990. Prior to joining the City he served as Director of the Alberta Main Street Program, a joint agreement between the Alberta Historical Resources Foundation and the Heritage Canada Foundation.

He is responsible for the development and implementation of the Stephen Avenue Heritage Area Program and was instrumental in establishing the agreement between Heritage Canada and the City of Calgary for the initiation of the Inglewood Main Street Program. Mr. Graham is the co-author of a study on Cultural Landscapes in Alberta for Alberta Community Development and has undertaken work for Parks Canada on the cultural landscape at Stirling Agricultural Village, National Historic Site and the town site of Field, B.C.

For the City of Calgary he is responsible for the development and implementation of the Calgary Heritage Authority, a merger of the Heritage Advisory Board and the Calgary Municipal Heritage Properties Authority, which will enable the City to address the challenges of heritage preservation in a climate of rapid economic expansion.

Jason Ness

Jason Ness holds undergraduate degrees in history and archaeology, and has a Master of Environmental Design (Planning), specializing in heritage planning, at the University of Calgary. He is also currently a director of the Calgary Civic Trust.

Larry Pearson

Larry Pearson is the Manager of the Protection and Stewardship Section of the Heritage Resource Management Branch of Alberta Community Development. Mr. Pearson has worked for Alberta Community Development for twenty-three years in a range of positions involving the development of Alberta's historic sites and the management of its heritage resources.

Mr. Pearson is a graduate of Queen's University where he obtained a BA (Hons) in Art History with a focus on architectural history. He has a Masters of Environmental Design in Architecture from the University of Calgary where his area of specialization was architectural preservation.

Mr. Pearson has been a member of the Association for Preservation Technology since 1979 and has served as a Director of that organization.

Mr. Pearson has received a number of Premier's Awards of Excellence for his work within the Alberta Government including a gold award for his role as Team Leader for the St. Onufrius Church project which saw a small Ukrainian Orthodox Church relocated from Smoky Lake Alberta to the History Hall of the Museum of Civilization in Hull, Quebec.

Panel Discussion Participants

Dr. Bill Bryne

Dr. Bryne was the founding director of the Alberta Archaeological Survey, and has been Assistant Deputy Minister of the Historical Resources Division since 1980. Since 1999 he has been Deputy Minister, Department of Community Development. He has presided over an incredible renaissance in Alberta heritage. He is one of the longest serving heritage figures in Canada.

Dr. Bryne received his Ph.D. from Yale University, focusing on Plains pottery. He has given innumerable papers on heritage and archaeology.

Don Douglas

Mr. Douglas is President & Chief Executive Officer of United Inc., a residential and multi-family housing and land developer in western Canada for over fifty years. United is a public company listed on the TSE.

Mr. Douglas serves as a director of a number of private and public and companies as well as not-for-profit foundations. He is also Chairman of the Board of the Calgary Centre for the Performing arts and a member of the World Presidents' Organization in Alberta.

Bob Holmes

Mr. Holmes is the Executive Officer responsible for planning and transportation at the City of Calgary. His responsibilities include all of the City's land use planning functions from formulation of policy to development approvals. He also has responsibility for the planning and operation of the City's roads and public transit system, including the infrastructure to support them. In the past year he has been responsible for developing and implementing the City's new ten year $1.2 billion Transportation Infrastructure Investment Program. This program is funded by a share of the Provincial Fuel Tax, contributions from the development industry, and City funds. He is currently the Chairman of the Management Committee for the Transportation Project Office – a public/private partnership that the City has formed with a consortium of private sector companies to manage the major projects (such as LRT and interchanges) in the City's Transportation Infrastructure Investment Program.

Mr. Holmes has been a Senior Executive with the City of Calgary since 1989 with responsibilities primarily in the planning, transportation, real estate, and community development areas. He has been the chairman of the Calgary Planning Commission since 1989. He has also served on the Board of Directors of the Calgary Convention Centre Authority, the Calgary Convention and Visitors Bureau, the Calgary Centre for Performing Arts, and the Calgary Transportation Authority.

From November 1997 to December 1998 Mr. Holmes was seconded to the Provincial Government as Deputy minister of Alberta Municipal Affairs.

Bob Holmes has an undergraduate degree from Queens University and a graduate degree in Community and Regional Planning from the University of British Columbia. He is a member of the Canadian Institute of Planners and the Urban Transportation Council of Canada.

Dr. Michael McMordie

Dr. McMordie has appointments in the Faculties of Environmental Design and was the Dean of General Studies at the University of Calgary from 1990 to 1998. Currently, Dr. McMordie serves as Director of the Resources and the Environment Program in the Faculty of Graduate Studies.

Dr. McMordie received a B.Arch from the University of Toronto in 1962 and a Ph.D. from Edinburgh University in 1972. He has worked in architectural practices in Toronto and Edinburgh and has taught at the Universities of Edinburgh and Calgary. He chaired Calgary's Ad Hoc Heritage Committee in the mid-seventies, and brought forward to City Council the proposals that resulted in the City's Heritage Planning Program. Dr. McMordie was the second President of the Society for the Study of Architecture in Canada. From the inception of the Canadian Encyclopaedia, he has advised on its architecture coverage, and has written many of its articles. He has also served as a consultant on historic architecture to the Cities of Calgary and Edmonton and to Parks Canada.

He also has affiliations with Clare Hall, Cambridge, the Society for the Study of Architecture in Canada, the Society of Architectural Historians of Great Britain, The University Art Association and the Society for the Study of Higher Education in Canada.

Harold Milavsky

Dr. Milavsky is the chairman of Quantico Capital Corp. and director on the boards of Aspen Properties Ltd., Citadel Diversified Management Ltd., Encal Energy Ltd., Northrock Resources Ltd., PrimeWest Energy Inc., TELUS Communications Inc., Torode Realty Limited and TransCanada PipeLines Limited. He has been a director for ENMAX since 1998.

Former President and CEO of Trizec Corporation Ltd., Harold Milavsky combines a distinguished business career with an impressive record of voluntarism and community service.

Dr. Milavsky is a former member of the Board of Governors of the University of Calgary. He was awarded the University of Calgary's highest academic honour, the Degree of Doctor of Laws in 1995 and also at the University of Saskatchewan in 1995.

Dr. Frits Pannekoek

Dr. Pannekoek is the Director of Information Resources at the University of Calgary and is also an Associate Professor in the Faculty of Communication and Culture. Since July 1998, his prime responsibility is to bring together information resources functions into a seamless service. These resources include the University Library (and its special collections, the Canadian Literary Archives and the Canadian Architectural Archives), the University Archives, the University of Calgary Press, and the Nickle Arts Museum.

Formerly, Dr. Pannekoek was the Director of Historic Sites Service, Alberta Community Development (1979–98) and was responsible for the Provincial Archives of Alberta (1992–96). Responsibilities included the planning and management of research, development and operation of historical resources in-situ, in Alberta.

Dr. Pannekoek's research interests lie in public and applied history. His most recent publication is "The Rise of a Heritage Priesthood" in *Preservation of What, for Whom? A Critical Look at Historical Significance* (Michael A. Tomlan, ed.). The National Council for Preservation Education, Ithaca, 1998.

Appendices

Appendix i

Alberta Historical Resources Act (Section 29)

Condition or covenant on land

29(1) A condition or covenant, relating to the preservation or restoration of any land or building, entered into by the owner of land and

(a) the Minister,
(b) the council of the municipality in which the land is located,
(c) the Foundation, or
(d) an historical organization that is approved by the Minister, may be registered with the Registrar of Land Titles.

(2) When a condition or covenant under subsection (1) is presented for registration, the Registrar of Land Titles shall endorse a memorandum of the condition or covenant on any certificate of title relating to that land.

(3) A condition or covenant registered under subsection (2) runs with the land and the person or organization under subsection (1) that entered into the condition or covenant with the owner may enforce it whether it is positive or negative in nature and notwithstanding that the person or organization does not have an interest in any land that would be accommodated or benefited by the condition or covenant.

(4) A condition or covenant registered under subsection (2) may be assigned by the person or organization that entered into it with the owner to any other person or organization mentioned in subsection (1), and the assignee may enforce the condition or covenant as if it were

the person or organization that entered into the condition or covenant with the owner.

(5) If the Minister considers it in the public interest to do so, the Minister may by order discharge or modify a condition or covenant registered under subsection (2), whether or not the Minister is a party to the condition or covenant.

(6) If the Minister discharges or modifies a condition or covenant under subsection (5), the Minister shall register a copy of the order with the Registrar of Land Titles and the Registrar of Land Titles shall endorse a memorandum discharging or modifying the condition or covenant on the certificate of title to the land.

(7) This section applies notwithstanding section 48 of the *Land Titles Act*.

(8) No condition or covenant under this section is deemed to be an encumbrance within the meaning of the *Land Titles Act*.

RSA 1980 cH-8 s25;1994 cM-26.1 s642(28);1996 c32 s5(34)

[now RSA 2000, cH-9 s29] [Reproduced by permission of Alberta Queen's Printer.]

Appendix ii

Calgary Civic Trust – Sample Covenant and Agreement

A Sample Agreement based on an agreement had between the Minister of Community Development and the Calgary Civic Trust dated _____, 2000.

Whereas it is in the public interest to preserve Calgary's heritage, and whereas the Minister has the authority to enter into an agreement with the Calgary Civic Trust, an Alberta registered not for profit historical organization, pursuant to Section 29 of the *Historical Resources Act*, the Minister and the Society agree as follows:

1. Pursuant to Section 29 of the Historical Resources Act the Minister agrees to appoint The Calgary Civic Trust as "approved" "historical organization" which can enter into a 'condition or covenant,' relating to the preservation or restoration of any land or building" with the owner of the building or land
2. The Minister agrees not to modify a condition or covenant registered under Section 29 (2) without due consultation with the Calgary Civic Trust. Where the unilateral modification by the Minister of the covenant diminishes the value of the "covenant" held by the Trust the Minister agrees to pay the Trust the difference between the value of the covenant and its altered value.
3. The Civic Trust is appointed only to enter into covenants pursuant to this agreement with owners whose properties are located within the boundaries of Calgary and the surrounding areas.
4. The Civic Trust will hold the Minister harmless from any legal action resulting from any covenants entered into by the Trust, provided such legal actions do not result from the Minister's alteration of an existing covenant entered into pursuant to Section 29 of the *Historical Resources Act*.
5. The Civic Trust may only enter into covenants pursuant to this agreement with:

- property owners whose properties are designated Provincial and Registered Resources under the Historical Resources Act
- property owners whose have applied for designation under the *Historical Resources Act* and meet the assessment criteria for Provincial and Registered historical resources guidelines (Appendix A) established by the Historic Sites Service, Alberta Community Development.
- property owners whose property would meet the criteria for Provincial or Registered ranking with the application of the Historic Sites Service guidelines (Appendix A) properties considered of municipal significance by the Calgary Heritage Advisory Board pursuant to the *Historical Resources Act*

The Calgary Civic Trust will give paramount consideration to the heritage assessments given to any property by the Historic Sites Service, Alberta Community Development where these are available before entering into a covenant pursuant to this agreement.

6. The Civic Trust will enter into an agreement with the Alberta Historical Resources Foundation or any other agency as may be directed by the Minister to assist in the management of the covenants. The Civic Trust agrees not to dispose of or alienate its covenants except to the Alberta Historical Resources Foundation or any other agency or society designated by the Minister. The Civic Trust also agrees to relinquish the covenants entered into pursuant to this agreement to the Alberta Historical Resources Foundation or other agency or society designated by the Minister should the Trust cease to exist, or become insolvent.
7. The Civic Trust will use Appendix B as the basic covenanting instrument. It will only be amended in substance with the approval of the Minister or his delegate.
8. The Civic Trust will use the Alberta Community Development's Guide to the Preservation and Rehabilitation of Historic Structures (Appendix C) for the management of its covenants and historic fabric.
9. The Civic Trust agrees not to amend its by laws without consultation with the Minister's or his designate.

10. The Civic Trust agrees to hold the Minister harmless from any legal action resulting from the Trust's management of the covenants.
11. The Civic Trust will be responsible for the costs emanating from the management of its covenants entered into pursuant to this agreement. This will however not preclude the Civic Trust or property owners who have entered into covenants from being eligible to apply for Alberta Historical Resource Foundation grants.
12. The Civic Trust will provide the Minister and the City of Calgary Heritage Advisory Board with an annual report on its activities and on the conditions of the buildings on which it holds covenants.
13. The Minister agrees not to appoint any other society or organization in Calgary pursuant to Section 29of the Act as covenanting agents so long as this agreement is in force without the approval of the Civic Trust.
14. The Minister agrees to make available the expertise of the Historic Sites Service in assisting in determining heritage value, in negotiating covenants, and in the dealings of the Calgary Civic Trust with federal government agencies particularly Revenue Canada, and National Historic Sites.
15. The Trust agrees not to enter into any agreement without first informing and securing advice from the Historic Sites Service.

Model Historic Preservation Easement

THIS PRESERVATION AND CONSERVATION EASEMENT DEED, made this _____ day of _____, 2002, by and between _____ _____ ("Grantor") and ("Grantee"), a non-profit corporation of [province of incorporation].
WITNESSETH:
WHEREAS, Grantor is owner in fee simple of certain real property located in the [town, county, and state], more particularly described in Exhibit A attached hereto and incorporated herein (hereinafter "the Property"), said Property including the following structures (hereinafter "the Buildings"):

the principal residence constructed of [brief description] dating from [year] (hereinafter "the Residence"); and additional ancillary structure [describe] (hereinafter "the Ancillary Structures").

[WHEREAS, the Property also includes a formal landscaped garden, [describe], designed by noted landscape architect [name] (hereinafter "the Garden");]

WHEREAS, the Property has significant undeveloped open space, including fields, forests, and [describe other], that contributes to the setting, context, and the public's view of the Buildings;

WHEREAS, Grantee is authorized to accept preservation and conservation Easements to protect property significant in national and state history and culture under the provisions of [*The Alberta Historical Resources Act*] (hereinafter "the Act");

WHEREAS, Grantee is a nonprofit organization whose primary purposes include the preservation and conservation of sites, buildings, and objects of significance to the City of Calgary and the Province of Alberta.

WHEREAS, the Property stands as a significant example of style architecture in Alberta and the City of Calgary illustrates aesthetics of design and setting, and possesses integrity of materials and workmanship;

WHEREAS, Grantor and Grantee recognize the architectural, historic, and cultural values (hereinafter "conservation and preservation values") and significance of the Property, and have the common purpose of conserving and preserving the aforesaid conservation and preservation values and significance of the Property;

WHEREAS, the Property's conservation and preservation values are documented in a set of reports, drawings, and photographs (hereinafter, "Baseline Documentation") incorporated herein by reference, which Baseline Documentation the parties agree provides an accurate representation of the Property as of the effective date of this grant. In the event of any discrepancy between the two counterparts produced, the counterpart retained by Grantee shall control;

WHEREAS, the Baseline Documentation shall consist of the following: [list documents and materials]

WHEREAS, the grant of a preservation and conservation Easement by Grantor to Grantee on the Property will assist in preserving and maintaining the Property and its architectural, historic, and cultural features for the benefit of the people of the City of Calgary and the Province of Alberta.

WHEREAS, to that end, Grantor desires to grant to Grantee, and Grantee desires to accept, a preservation and conservation Easement (hereinafter, the "Easement") in gross in perpetuity on the Property present to the Act.

NOW, THEREFORE, pursuant to Section 29of the Alberta Historical Resources Division, the Grantor does hereby voluntarily grant and convey unto the Grantee a preservation and conservation Easement in gross in perpetuity over the Property described in Exhibit A.

Purpose

1.　**Purpose.** It is the Purpose of this Easement of assure that the architectural, historic, and cultural aspects of the building will be retained and maintained forever in close to its original condition, subject to upgrade to meet current building codes. Such upgrades will be acceptable to the Alberta Historic Site consultant and will not significantly alter the outer or inner historic details of the Building. Once restored and upgraded, the Easement will ensure that the Property will be retained and maintained forever substantially in its restored condition for conservation and preservation purposes and to prevent any use or change of the Building that will significantly impair or interfere with the Building's conservation and preservation values.

Grantor's Covenants

2.1　**Grantor's Covenants: Covenant to Maintain.** Grantor agrees at all times to maintain the Buildings in the same structural condition and state of repair as that existing on completion of the restoration, on a date to be established by Grantor and Grantee. Grantor's obligation to maintain shall require replacement, repair, and reconstruction by Grantor whenever necessary to preserve the Building in substantially the same structural condition and state of repair as that existing on the date specified in this Easement, on completion of the restoration. Grantor's obligation to maintain shall also require that the Property's landscaping be maintained in good appearance with substantially similar plantings, vegetation, and natural screening to that existing between [the relevant years] around the Building. The existing lawn areas shall be maintained as lawns, regularly mown. The existing meadows and open fields shall be maintained as meadows and open fields, regularly bush-hogged to prevent the growth of woody vegetation where none existed. Subject to the casualty provisions of Paragraphs 7 and 8, this obligation to maintain shall require replacement, rebuilding, repair, and reconstruction of the Buildings whenever necessary in accordance with The Historic Sites Service's Standards for Rehabilitation and Guidelines for Rehabilitating Historic Buildings, as these may be amended from time to time.

2.2 **Grantor's Covenants: Prohibited Activities.** The following acts or uses are expressly forbidden on, over, or under the Property, except as otherwise condition in this paragraph:

a) the Buildings shall not be demolished, removed, or razed except as provided in paragraphs 7 and 8;
b) nothing shall be erected or allowed to grow on the Property which would impair the visibility of the Property and the Buildings from street level;
c) no other buildings or structures, including satellite receiving dishes (small rooftop dishes excluded), camping accommodations, or mobile homes, shall be erected or placed on the Property hereafter except for temporary structures required for the maintenance or rehabilitation of the Property, such as construction trailers;
d) the dumping of ashes, trash, rubbish, or any other unsightly or offensive materials is prohibited on the Property:
e) the Property shall not be divided or subdivided in law or in fact and the Property shall not be devised or conveyed except as a unit;
f) no above-ground utility transmission lines, except those reasonably necessary for the existing Buildings, may be created on the Property, subject to utility Easements already recorded;

g) subject to the maintenance covenants of paragraph 2.1 hereof, the following features located within the Residence [or Buildings/Ancillary Structures] shall not be removed, demolished, or altered;

h) the main floor of the Building shall be restored to and maintained in near original condition as of [the relevant years], subject to building code upgrades required to make the Building usable by the public.

Grantor's Conditional Rights

3.1   **Conditional Rights Requiring Approval by Grantee.** Once the Building is moved to its new site and restored under the guidance of the consultant from the Historic Sites Service, without the prior express written approval of the Grantee, which approval may be withheld or conditioned in the sole discretion of Grantee, Grantor shall not undertake any of the following actions:

a) increase or decrease the height of, make additions to, change the exterior construction materials or colours of, or move, improve, alter, reconstruct, or change the facades (including fenestration) and roofs of the Buildings;

b) change the floor plan of the School;

c) erect any external signs or external advertisements except: (i) such plaque permitted under paragraph 19 of this Easement; (ii) a sign stating solely the address of the Property; and (iii) a temporary sign to advertise the sale or rental of the Property;

d) make permanent substantial topographical changes, such as, by example, excavation for the construction of roads and recreational facilities;

e) cut down or otherwise remove live trees located within existing lawn areas, or cut down or otherwise remove live trees located outside the existing lawn areas, meadows and open fields for the purpose of conducting commercial timber production [or allow conditional harvesting of timber in accordance with qualified plan presented to Grantee for approval. (See Model Conservation Easement)]; and

f) change the use of the Property to another use other than [single family residential]. Grantee must determine that the proposed use: (i) does not impair the significant conservation and preservation values of the Property; and (ii) does not conflict with the Purpose of the Easement.

3.2 Review of Grantor's Requests for Approval. Grantor shall submit to Grantee for Grantee's approval of those conditional rights set out at paragraph 3.1 two copies of information (including plans, specifications, and designs where appropriate) identifying the proposed activity with reasonable specificity. In connection therewith, Grantor shall also submit to Grantee a timetable for the proposed activity sufficient to permit Grantee to monitor such activity. Within 45 (forty-five) days of Grantee's receipt of any plan or written request for approval hereunder, Grantee shall certify in writing that (a) it approves the plan or request, or (b) it disapproves the plan or request as submitted, in which case Grantee shall provide Grantor with written suggestions for modification or a written explanation for Grantee's disapproval. Any failure by Grantee to act within 45 (forty-five) days of receipt of Grantor's submission or resubmission of plans or requests shall be deemed to constitute approval by Grantee of the plan or request as submitted and to permit Grantor to undertake the proposed activity in accordance with the plan or request submitted.

4. Standards for Review. In exercising any authority created by the Easement to inspect the Property or the interior of the Residence; to review any construction, alteration, repair, or maintenance; or to review casualty damage or to reconstruct or approve reconstruction of the Building following casualty damage, Grantee shall apply the Provinces Standards.

5. Public Access. At times deemed reasonable by Grantor person affiliated with educational organizations, professional architectural associations, and historical societies shall be admitted to study the property. Grantee may make photographs, drawings, or other representations documenting the significant historical, cultural, and architectural character and features of the property and distribute them to magazines, newsletters, or other publicly available publications, or use them to fulfill its charitable and educational purposes.

Grantor's Reserved Rights

6.   **Grantor's Reserved Rights Not Requiring Further Approval by Grantee.** Subject to the provisions of paragraphs 2.1, 2.2, and 3.1, the following rights, uses, and activities of or by Grantor on, over, or under the Property are permitted by this Easement and by Grantee without further approval by Grantee:

- a) the right to engage in all those acts and uses that: (i) are permitted by governmental statute or regulation; (ii) do no substantially impair the conservation and preservation values of the Property; and (iii) are not inconsistent with the Purpose of this Easement;
- b) pursuant to the provisions of paragraph 2.1, the right to maintain and repair the Buildings strictly according to the Province's Standards. As used in this subparagraph, the right to maintain and repair shall mean the use by Grantor of in-kind materials and colours, applied with workmanship comparable to that which was used in the construction or application of those materials being repaired or maintained, for the purpose of retaining in good condition the appearance and construction of the Buildings. The right to maintain and repair as used in this subparagraph shall not include the right to make changes in appearance, materials, colours, and workmanship from that existing prior to the maintenance and repair without the prior approval of Grantee in accordance with the provisions of paragraphs 3.1 and 3.2;
- c) the right to continue all manner of existing residential use and enjoyment of the Property's Buildings and Garden, including but not limited to the maintenance, repair, and restoration of existing fences; the right to maintain existing driveways, roads, and paths with the use of same or similar surface materials; the right to maintain existing utility lines, gardening and building walkways, steps, and garden fences; the right to cut, remove, and clear grass or

other vegetation and to perform routine maintenance, landscaping, horticultural activities, and upkeep, consistent with the Purpose of this Easement; and

d) the right to conduct at or on the Property educational and nonprofit activities that are not inconsistent with the protection of the conservation and preservation values of the Property.

Casualty Damage or Destruction; Insurance

7. **Casualty Damage or Destruction.** In the event that the Buildings or any part thereof shall be damaged or destroyed by fire, flood, windstorm, hurricane, earth movement, or other casualty, Grantor shall notify Grantee in writing within fourteen (14) days of the damage or destruction, such notification including what, if any, emergency work has already been completed. No repairs or reconstruction of any type, other than temporary emergency work to prevent further damage to the Buildings and to protect public safety, shall be undertaken by Grantor without Grantee's prior written approval. Within thirty (30) days of the date of damage or destruction, if required by Grantee, Grantor at its expense shall submit to the Grantee a written report prepared by a qualified restoration architect and an engineer who are acceptable to Grantor and Grantee, which report shall include the following;

a) an assessment of the nature and extent to the damage;
b) a determination of the feasibility of the restoration of the Buildings and/or reconstruction of damaged or destroyed portions of the Buildings; and
c) a report of such restoration/reconstruction work necessary to return the Buildings to the condition existing at the date hereof.

8. **Review After Casualty Damage or Destruction.** If, after reviewing the report provided in paragraph 7 and assessing the availability of insurance proceeds after satisfaction of any mortgagee's/lender's claim under paragraph 9, Grantor and Grantee agree that the Purpose of the Easement will be served by such restoration/reconstruction, Grantor and Grantee shall establish a schedule under which Grantor shall complete the restoration/reconstruction

of the Buildings in accordance with plans and specifications consented to by the parties up to at least the total of casualty insurance proceeds available to Grantor.

If, after reviewing the report and assessing the availability of insurance proceeds after satisfaction of any mortgagee's/lender's claims under paragraph 9, Grantor and agree that restoration/reconstruction of the Property is impractical or impossible, or agree that the Purpose of the Easement would not be served by such restoration/reconstruction, Grantor may, with the prior written consent of Grantee, alter, demolish, remove, or raze one or more of the Buildings, and/or construct new improvements on the Property. Grantor and Grantee may agree to extinguish this Easement in whole or in part in accordance with the laws of the Province of and paragraph 23.2 hereof.

If, after reviewing the report and assessing the availability of insurance proceeds after satisfaction of any mortgagee's/lender's claims under paragraph 9, Grantor and Grantee are unable to agree that the Purpose of the Easement will or will not be served by such restoration/reconstruction, the matter may be referred by either party to binding arbitration and settled in accordance with Province of Alberta's arbitration statute then in effect.

9.   **Insurance.** Grantor shall keep the Property insured against loss from the perils commonly insured under standard fire and extended coverage policies and comprehensive general liability insurance against claims for personal injury, death, and property damage. Property damage insurance shall include change in condition and building ordinance coverage, in form and amount sufficient to replace fully the damaged Property and Buildings without cost or expense to Grantor or contribution or coinsurance from Grantor. Such insurance shall include grantee's interest and name Grantee as an additional insured. Grantor shall deliver to Grantee, within ten (10) business days of Grantee's written request therefore, certificates of such insurance coverage. Provided, however, that whenever the Property is encumbered with a mortgage or deed of trust, nothing contained in this paragraph shall jeopardize the prior claim, if any, of the mortgagee/lender to the insurance proceeds.

Indemnification; Taxes

10. **Indemnification.** Grantor hereby agrees to pay, protect, indemnify, hold harmless and defend at its own cost and expense, Grantee, its agents, directors, officers and employees, or independent contractors from and against any and all claims, liabilities, expenses, costs, damages, losses, and expenditures (including reasonable attorneys' fees and disbursements hereafter incurred) arising out of or in connection with injury to or death of any person; physical damage to the Property, the presence or release in, on, or about the Property, at any time, of any substance now or hereafter defined, listed, or otherwise classified pursuant to any law, ordinance, or regulation as a hazardous, toxic, polluting, or contaminating substance; or other injury or other damage occurring on or about the Property, unless such injury or damage is caused by Grantee or any agent, director, officer, employee, or independent contractor of Grantee. In the event that Grantor is required to indemnify Grantee pursuant to the terms of this paragraph, the amount of such indemnity, until discharged, shall constitute a lien on the Property with the same effect and priority as a mechanic's lien. Provided, however, that nothing contained herein shall jeopardize the priority of any recorded lien of mortgage or deed of trust given in connection with a promissory note secured by the Property.

11. **Taxes.** Grantor shall pay immediately, when first due and owing, all general taxes, special taxes, special assessments, water charges, sewer service charges, and other charges which may become a lien on the Property unless Grantor timely objects to the amount or validity of the assessment or charge and diligently prosecutes an appeal thereof, in which case the obligation hereunder to pay such charges shall be suspended for the period permitted by law for prosecuting such appeal and any applicable grace period following completion of such action. In place of Grantor, Grantee is hereby authorized, but in no event required or expected, to make or advance upon three (3) days prior written notice to Grantor any payment relating to taxes, assessments, water rates, sewer rentals and other governmental or municipality charge, fine, imposition, or lien asserted against the Property. Grantee may make such payment according to any bill, statement, or estimate procured from the appropriate public office without inquiry into the accuracy of such bill,

statement, or assessment or into the validity of such tax, assessment, sale, or forfeiture. Such payment if made by Grantee shall constitute a lien on the Property with the same effect and priority as a mechanic's lien, except that such lien shall not jeopardize the priority of any recorded lien of mortgage or deed of trust given in connection with a promissory note secured by the Property.

Administration and Enforcement

12. **Written Notice.** Any notice which either Grantor or Grantee may desire or be required to give to the other party shall be in writing and shall be delivered by one of the following methods – by overnight courier postage prepaid, facsimile transmission, registered or certified mail with return receipt requested, or hand delivery; if to Grantor, then at [address], and if to Grantee, then to [address].
Each party may change its address set forth herein by a notice to such effect to the other party.

13. **Evidence of Compliance.** Upon request by Grantor, Grantee shall promptly furnish Grantor with certification that, to the best of Grantee's knowledge, Grantor is in compliance with the obligations of Grantor contained herein or that otherwise evidences the status of this Easement to the extent of Grantee's knowledge thereof.

14. **Inspection.** With the consent of Grantor, representatives of Grantee shall be permitted at all reasonable times to inspect the Property, including the interior of the Residence [or Buildings/Ancillary Structures]. Grantor covenants not to withhold unreasonably its consent in determining dates and times for such inspections.

15. **Grantee's Remedies.** Grantee may, following reasonable written notice to Grantor, institute suit(s) to enjoin any violation of the terms of this Easement by ex parte, temporary, preliminary, and/or permanent injunction, including prohibitory and/or mandatory injunctive relief and to require the restoration of the Property and Buildings to the condition and appearance

that existed prior to the violation complained of. Grantee shall also have available all legal and other equitable remedies to enforce Grantor's obligations hereunder.

In the event Grantor is found to have violated any of its obligations, Grantor shall reimburse Grantee for any costs or expenses incurred in connection with Grantee's enforcement of the terms of this Easement, including all reasonable court costs, and attorney's, architectural, engineering, and expert witness fees.

Exercise by Grantee of one remedy hereunder shall not have the effect of waiving or limiting any other remedy, and the failure to exercise any remedy shall not have the effect of waiving or limiting the use of any other remedy or the use of such remedy at any other time.

16.   Notice from Government Authorities. Grantor shall deliver to Grantee copies of any notice of violation or lien relating to the Property received by Grantor from any government authority within five (5) days of receipt by Grantor. Upon request by Grantee, Grantor shall promptly furnish Grantee with evidence of Grantor's compliance with such notice or lien where compliance is required by law.

17.   Notice of Proposed Sale. Grantor shall promptly notify Grantee in writing of any proposed sale of the Property and provide the opportunity for Grantee to explain the terms of the Easement to potential new owners prior to sale closing.

18.   Liens. Any lien on the Property created pursuant to any paragraph of this Easement may be confirmed by judgement and foreclosed by Grantee in the same manner as a mechanic's lien, except that now lien created pursuant to this Easement shall jeopardize the priority of any recorded lien of mortgage or deed of trust given in connection with a promissory note secured by the Property.

19.   Plaque. Grantor agrees that Grantee may provide and maintain a plaque on the Property, which plaque shall not exceed 24 by 24 inches in size, giving notice of the significance of the [The property].

Binding Effect; Assignment

20.   **Runs with the Land.** Except as provided in paragraphs 8 and 23.2, the obligations imposed by this Easement shall be effective in perpetuity and shall be deemed to run as a binding servitude with the Property. This Easement shall extend to and be binding upon Grantor and Grantee, their respective successors in interest and all persons hereafter claiming under or through Grantor and Grantee, and the words "Grantor" and "Grantee" when used herein shall include all such persons. Any right, title, or interest herein granted to Grantee also shall be deemed granted to each successor and assign of Grantee and each such following successor and assign thereof, and the word "Grantee" shall include all such successors and assigns.

Anything contained herein to the contrary notwithstanding, an owner of the Property shall have no obligation pursuant to this instrument where such owner shall cease to have any ownership interest in the Property by reason of a bona fide transfer. The restrictions, stipulations, and covenants contained in this Easement shall be inserted by Grantor, verbatim or by express reference, in any subsequent deed or other legal instrument by which Grantor divests itself of either the fee simple title to or any lesser estate in the Property or any part thereof, including by way of example and not limitation, a lease of all or a portion of the Property.

21.   **Assignment.** Grantee may convey, assign, or transfer this Easement to the provincial government or to a similar civic, provincial, or national organization whose purposes are to promote preservation or conservation of historical, cultural, or architectural resources, provided that any such conveyance, assignment, or transfer requires that the Purpose for which the Easement was granted will continue to be carried out.

22.   **Recording and Effective Date.** Grantee shall do and perform at its own cost all acts necessary to the prompt recording of this instrument in the land records of the Province of Alberta. Grantor and Grantee intend that the restrictions arising under this Easement take effect on the day and year this instrument is recorded in the land records.

Percentage Interests; Extinguishment

23.1   Percentage Interests. For purposes of allocating proceeds pursuant to paragraphs 23.2 and 23.3. Grantor and Grantee stipulate that as of the date of this Easement, Grantor and Grantee are each vested with real property interests in the Property and that such interests have a stipulated percentage interest in the fair market value of the Property. Said percentage interests shall be determined by the ratio of the value of the Easement on the effective date of this Easement to the value of the Property, without deduction for the value of the Easement, on the effective date of this Easement. The values on the effective date of the Easement shall be those values used to calculate the deduction for federal income tax purposes. The parties shall include the ratio of those values with the Baseline Documentation (on file with Grantor and Grantee) and shall amend such values, if necessary, to reflect any final determination thereof by Revenue Canada or court of competent jurisdiction. For purposes of this paragraph, the ratio of the value of the Easement to the value of the Property unencumbered by the Easement shall remain constant, and the percentage interests of Grantor and Grantee in the fair market value of the Property thereby determinable shall remain constant, except that the value of any improvements made by Grantor after the effective date of this Easement is reserved to Grantor.

23.2   Extinguishment. Grantor and Grantee hereby recognize that circumstances may arise that may make impossible the continued ownership or use of the Property in a manner consistent with the Purpose of this Easement and necessitate extinguishment of the Easement. Such circumstances may include, but are not limited to, partial or total destruction of the buildings resulting from casualty. Extinguishment must be the result of a judicial proceeding in a court of competent jurisdiction. Unless otherwise required by applicable law at the time, in the event of any sale of all or a portion of the Property (or any other property received in connection with an exchange or involuntary conversion of the Property) after such termination or extinguishment, and after the satisfaction of prior claims and any costs or expenses associated with such sale, Grantor and Grantee shall share in any net proceeds resulting from such sale in accordance with their respective percentage interests in the fair market value of the Property, as such interests are determined under the provisions of paragraph 23.1, adjusted, if necessary, to reflect a partial termination or extinguishment of this Easement. All such proceeds received by Grantee shall be used by Grantee in a manner consistent with Grantee's primary purposes. Net proceeds shall also include, without limitation, net insurance proceeds.

In the event of extinguishment, the provisions of this paragraph shall survive extinguishment and shall constitute a lien on the Property with the same effect and priority as a mechanic's lien, except that such lien shall not jeopardize the priority of any recorded lien of mortgage or deed of trust given in connection with a promissory note secured by the Property.

23.3 Condemnation. If all or any part of the property is taken under the power of eminent domain by public, corporate, or other authority, or otherwise acquired by such authority through a purchase in lieu of a taking, Grantor and Grantee shall join in appropriate proceedings at the time of such taking to recover the full value of those interests in the Property that are subject to the taking and all incidental and direct damages resulting from the taking. After the satisfaction of prior claims and net of expenses reasonably incurred by Grantor and Grantee in connection with such taking, Grantor and Grantee shall be respectively entitled to compensation from the balance of the recovered proceeds in conformity with the provisions of paragraphs 23.1 and 23.2 unless otherwise provided by law.

Interpretation

24. Interpretation. The following provisions shall govern the effectiveness, interpretation, and duration of the Easement

- a) Any rule of strict construction designed to limit the breadth of restrictions on alienation or use of Property shall not apply in the construction or interpretation of this Easement, and this instrument shall be interpreted broadly to effect its Purpose and the transfer of rights and the restrictions on use herein contained.
- b) This instrument may be executed in two counterparts, one of which may be retained by Grantor and the other, after recording,

to be retained by Grantee. In the event of any disparity between the counterparts produced, the recorded counterpart shall in all cases govern.

c) This instrument is made pursuant to the Act, but the invalidity of such Act or any part thereof shall not affect the validity and enforceability of this Easement according to its terms, it being the intent of the parties to agree and to bind themselves, their successors, and their assigns in perpetuity to each term of this instrument whether this instrument be enforceable by reason of any statute, common law, or private agreement in existence either now or hereafter. The invalidity or unenforceability of any provision of this instrument shall not affect the validity or enforceability of any other provision of this instrument or any ancillary or supplementary agreement relating to the subject matter thereof.

d) Nothing contained herein shall be interpreted to authorize or permit Grantor to violate any ordinance or regulation relating to building materials, construction methods, or use. In the event of any conflict between any such ordinance or regulation and the terms hereof, Grantor promptly shall notify Grantee of such conflict and shall cooperate with Grantee and the applicable governmental entity to accommodate the purposes of both this Easement and such ordinance or regulation.

e) To the extent that Grantor owns or is entitled to development rights which may exist now or at some time hereafter by reason of the fact that under any applicable zoning or similar ordinance the Property may be developed to use more intensive (in terms of height, bulk, or other objective criteria related by such ordinances) than the Property is devoted as of the date hereof, such development rights shall not be excisable on, above, or below the Property during the terms of the Easement, nor shall they be transferred to any adjacent parcel and exercised in a manner that would interfere with the Purpose of the Easement.

Amendment

25.   **Amendment.** If circumstances arise under which an amendment to or modification of this Easement would be appropriate, Grantor and Grantee may by mutual written agreement jointly amend this Easement, provided that no amendment shall be made that will adversely affect the qualification of this Easement or the status of Grantee under any applicable laws of the Province of Alberta. Any such amendment shall be consistent with the protection of the conservation and preservation values of the Property and the Purpose of this Easement; shall not affect its perpetual duration; shall not permit additional residential development on the Property other than the residential development permitted by this Easement on its effective date; shall not permit any private inurement to any person or entity; and shall not adversely impact the overall architectural, historic, natural habitat, and open space values protected by this Easement. Any such amendment shall be recorded in the land records of Alberta. Nothing in this paragraph shall require Grantor or Grantee to agree to any amendment or to consult or negotiate regarding any amendment.

THIS EASEMENT reflects the entire agreement of Grantor and Grantee. Any prior or simultaneous correspondence, understandings, agreements, and representations are null and void upon execution hereof, unless set out in this instrument.

Mortgage Subordination [as applicable]

26.   **Subordination of Mortgage.** At the time of the conveyance of this Easement, the Property is subject to a Mortgage/Deed of Trust dated _____, recorded in the Land Records Alberta (hereinafter "the Mortgage"/ "the Deed of Trust") held by (hereinafter, "Mortgagee"/"Lender"). The Mortgagee/Lender joins in the execution of this Easement to evidence its agreement to subordinate the Mortgage/the Deed of Trust to this Easement under the following conditions and stipulations:

a) The Mortgagee/Lender and its assignees shall have a prior claim to all insurance proceeds as a result of any casualty, hazard, or accident occurring to or about the Property and all proceeds of condemnation proceedings, and shall be entitled to same in preference to Grantee until the Mortgage/the Deed of trust is paid off and discharged, notwithstanding that the Mortgage/the Deed of Trust is paid off and discharged, notwithstanding that the Mortgage/the Deed of Trust is subordinate in priority to the Easement.

b) If the Mortgagee/Lender receives an assignment of the leases, rents, and profits of the Property as security or additional security for the loan secured by the Mortgage/Deed of Trust, then the Mortgagee/Lender shall have a prior claim to the leases, rents, and profits of the Property and shall be entitled to receive same in preference to Grantee until the Mortgagee's/Lender's debt is paid off or otherwise satisfied, notwithstanding that the Mortgage/Deed of Trust is subordinate in priority to the Easement.

c) The Mortgagee/Lender or purchaser in foreclosure shall have no obligation, debt, or liability under the Easement until the Mortgagee/Lender or a purchaser in foreclosure under it obtains ownership of the Property. In the event of foreclosure or deed in lieu of foreclosure, the Easement is not extinguished.

d) Nothing contained in this paragraph or in this Easement shall be construed to give any Mortgage/Lender the right to violate the terms of this Easement or to extinguish this Easement by taking title to the Property by foreclosure or otherwise.

[Signature]TO HAVE AND TO HOLD, the said Preservation and Conservation Easement, unto the said Grantee and its successors and permitted assigns forever. This DEED OF PRESERVATION AND CONSERVATION EASEMENT may be executed in two counterparts and by each party on a separate counterpart, each of which when so executed and delivered shall be an original, but both of which together shall constitute one instrument. IN WITNESS WHEREOF, Grantor and Grantee have set this hands under seal on the days and year set forth below.

WITNESS:		GRANTOR:		(date):
ATTEST:		GRANTEE:
By:			By:
Its President:
[Notarization]					(date):

Appendix iii

Web Sites for Covenanting Information

Name of Organization	Web Sites Verified (2004-04-7)
Alberta Real Estate Legal Questions	http://www.law-faqs.org/ab/real-gen.htm
American Farm Bureau, US.	http://www.fb.com/
Arkansas Historic Preservation Program, AR, US.	http://www.arkansaspreservation.org/
California Land Title Association Legal Developments, CA, US.	http://www.clta.org/
Canada Business Service Centre, CA.	http://www.cbsc.org/english
Canadian Real Estate Law, CA.	http://www.canadalegal.com/gosite.asp
Charlotte-Meklenburg Historic Preservation Foundation, NC, US.	http://www.cmhpf.org/
Delta Water Fowl Foundation, Public Policy, US.	http://www.deltawaterfowl.org/
Franklin Massachusetts, Implementing actions, MA, US.	http://www.franklin.ma.us/town/masterplan/
Historic Charleston Foundation, SC, US.	http://www.historiccharleston.org/easements.html
The Island Trust Fund, BC, CA	http://www.islandstrustfund.bc.ca/
Land Trust Alliance e-mail mailing list, US.	http://www.onelist.com/subscribe.cgi/landstrust
Land Trust Alliance, US.	http://www.lta.org/
Law Society Journal, Australia	http://www.lawsociety.com.au/
Leelanau Conservancy, MI, US.	http://www.theconservancy.com/

Minnesota Land Trust (handbook), MN, US.	http://www.mnland.org/
Ministry of Consumer and Business Service Canada, CA.	http://www.cbs.gov.on.ca/mcbs/english/welcome.htm
Montana Land Reliance, MT, US.	http://www.mtlandreliance.org/
Montana State University Land Use Articles, MT, US.	http://www.montana.edu/wwwpb/reso/reso_idx.html
Office of the Deputy Under Secretary of Defense, US. (Institutional Controls)	http://www.dtic.mil/
Ontario Heritage Foundation, ON, CA.	http://www.heritagefdn.on.ca/Eng/home-eng.shtml
Ontario Ministry of Culture	http://www.culture.gov.on.ca/
Pace Law School: Preserving Open Space with Land Trusts and Conservation Easements, NY, US.	http://www.law.pace.edu/landuse/lndtrs.html
Real Estate Centre, US.	http://recenter.tamu.edu/news
Revenue Canada, CA.	http://www.ccra-adrc.gc.ca/
Society for the Preservation of New England Antiquities, MA, US.	http://spnea.org/
Southern Appalachian Highland Conservatory, US.	http://www.appalachian.org/
Southwestern Archaeology, US.	http://www.swanet.org/
Stephen J. Small, attorney, private land protection options and strategies, US.	http://www.stevesmall.com/
Strategis: Canada's Business and Consumer Site, CA.	http://www.strategis.ic.gc.ca/
University of Pittsburgh, School of Law (Resources in Property Law), PA, US.	http://www.law.pitt.edu/library
Victoria Law Institute, Australia	http://www.liv.asn.au/
West Coast Environmental Law Research Foundation, BC, CA.	http://www.vcn.bc.ca/
West Legal Directory, Canada and US.	http://wireless.wld.com/wl/index.jsp

Appendix iv

List of Resources for Covenanting

Barrett, Thos. S. The Conservation Easement in California.

Barrett, Thos. S. Land Trust Exchange and Trust for Public Land, San Francisco.

CAGP.ACPOP. Guide to Estate Planning and Charitable Giving. Canadian Association of Gift Planners.

Conserving the Rural Landscape: A Symposium. Environmental Design, U of Calgary, 1992.

Curthoys, Lesley P. For the Love of Alberta – ways to save natural heritage. Private Conservancy Guide for Alberta.

Denhez, Marc. The Heritage Strategy Planning Handbook. Dundern Press, 1997.

Diehl, Janet. The Conservation Easement Handbook.

Duerksen, Chris. J. A Handbook on Historic Preservation Law, 1983, Washington DC 20036.

Easements and Restoration Covenants, Programme II 726. Law Society of Upper Canada, Toronto, 1989.

Gammage, Grady, Jr., Philip N. Jones, and Stephen L. Jones. Historic Preservation in California: A Legal Handbook. Stanford, CA: Stanford Environmental Law Society. 1975.

King, Thos. F. Cultural Resource Laws and Practice – A Guide. Alta Mira Press.

Kwasniak, Arlene J. "Facilitating Conservation: Private Conservancy Law Reform" Alberta Law Review 31, no. 4 (Nov. 1993): 607–23.

Kwasniak, Arlene J. Conservation Guide for Alberta. Environmental Law Centre, Edmonton, 1997.

Lewison, Kim & Pugh-Smith, John, eds. Taxation Incentives for Heritage PreservationProperty, Planning and Compensation Reports 1994, Vol. 67. Sweet & Maxwell, London, 1994.

Livermore, Putnam. Trust for Public Land. Island Press, Covelo, California, 1983.

Loukidelis, David. Using Conservation Covenants to Preserve Private Land in British Columbia West Coast Environmental Law Research Foundation, Vancouver, 1992.

Luxenburg, Gretchen. Ebey's Landing National History Reserve – Case Study of a Preserved Cultural Landscape.

Payne, Michael. Preserving and Interpreting Cultural Landscapes in Alberta. Alberta Community Development.

Siegel, Stephen A. A Student's Guide to Easements, Real Covenants and Equitable Servitudes Lerner Law Books, 2d ed., 1999. www.lawbooksUSA.com.

Small, Stephen J. Federal Tax Law of Conservation Easements, Land Trust Exchange, 1986.

Small, Stephen J. Preserving Family Lands Land Trust Exchange, 1988.

Ziegler Jr., Arthur P. Historic Preservation in Inner City Areas – Manual of Practice. Allegheny Press, 1971.

Ziff, Bruce. "Positive Covenants Running with the Land," Alberta Law Review 27, no. 3 (1989): 354–72.

Bibliography

Excellent bibliographic support can be found in:

Homik, Teresa M. (1985). "Heritage and the Alberta Planning Act," Historic Sites Services, Edmonton, Alberta.

> This has an extremely thorough analysis of heritage and planning issues particularly in the United States and Great Britain.

Witlib, Derek. (1993). "The Conservation Easement as a Stewardship Tool for Ontario," Senior Honours Paper, School of Urban and Regional Planning, University of Waterloo.

> It contains an appropriate literature review that those interested in covenants will find very useful.

Fitton, Michael E., John P. Hamilton, Garth Manning, and Arnold Weinrib (1989). "Easements and Restrictive Covenants," Toronto: Department of Education The Law Society of Upper Canada.

> Everyone will find this book an extremely useful self-educational tool. It has excellent citations and would be relevant to anyone interested in a full understanding of the issues.

Hodge, Ian, Richard Castle and Janet Dwyer (1993) "Covenants as a Conservation Mechanism," Land Economy Monograph 26. Cambridge: Granta Editions Limited.

> It contains an exhaustive bibliography of the British experience.

http://www.lta.org/pubs.html has an interesting collection of books videos and brochures on American land protection issues. There are a lot of simple "how to" videos and fund raising for convent volumes here.

The more recent materials are best accessed from the not for profit organizations and their WEB addresses listed in a separate section. These organizations often post up-to-date bibliographies. Legal on line materials are also a good source.

Notes

1 Alan F. J. Artibise and Jean Friesen, eds., "Heritage Conservation," Special Issue of *Prairie Forum* 15, no. 2 (fall 1990). This volume focuses on the prairie provinces and is probably the best single volume encompassing heritage preservation. For the only heritage bibliography compiled on Western Canada see F. Pannekoek, "A Selected Western Canadian Historical Resources Bibliography to 1985," *Prairie Forum* 15, no. 2 (fall 1990), pp. 329–74. It has not yet been updated.
2 http://www.heritagecanada.org/eng/main.html viewed July 14, 2003.
3 Frits Pannekoek, "The Rise of a Heritage Priesthood," in Michael A Tomlan, ed., *Preservation of What, For Whom?* (Ithaca: The National Council for Preservation Education, 1999), pp. 29–36.
4 Section 17 of the Ancient Monuments and Archaeological Areas Act, 1979, empowers certain central and local government bodies to enter into agreements with landowners governing the management of heritage sites. Those agreements may be made binding on subsequent owners and may contain positive, as well as negative, obligations. The provision is sometimes used by *English Heritage*, as an alternative to its other statutory powers, but I have not myself come across the provision being widely used outside that context. Unlike English Heritage (which is partly a government body), The National Trust (which is an independent charity) cannot rely on this provision.
5 This definition is taken from *Black's Law Dictionary*, 6th ed. (St. Paul, MN: West, 1990). It is further expanded in that dictionary as follows: "Traditionally the permitted kinds of uses were limited, the most important being rights-of-way and rights concerning flowing waters. The easement was normally for the benefit of adjoining lands, no matter who the owner was (an easement appurtenant), rather than for the benefit of a specific individual (easement in gross). The land, having the right of use as an appurtenance is known as the dominant tenant and the land which is subject to the easement is known as the servient tenement."

6   In 2002, the *Ontario Heritage Act* was amended to empower municipalities to prohibit demolition of designated properties until the owner has obtained a permit for a replacement building; the Act also requires that the replacement building be built within two years. In addition, the maximum fine for illegally demolishing designated properties was increased from $250,000 to $1 million.

7   In 2003, the Foundation's built heritage easement form was revised and differs significantly in form and content from the earlier form. The revised easement is oriented around a heritage character statement and heritage character defining features. It also includes guiding principles in conserving historic properties.

8   Alberta Land Trust Society, *Preserving Working Ranches in the Canadian West*

9   Canada Customs and Revenue Agency, *Gifts and Income Tax*, 2000.

10 Environment Canada, *Guidelines for Valuers*.

11 Environment Canada, *Canada's Ecological Gifts Program – Fact Sheet*.

12 Appraisal Institute of Canada and Environment Canada, *Ecological Gifts Seminar: Calgary, Alberta*, "Methodology," 2001.

13 Environment Canada, *The Ecological Gifts Program: Guidelines for Valuers*, 2001; Environment Canada, *The Ecological Gifts Program: The Appraisal Review and Determination Process*, draft, 2001.

14 Appraisal Institute of Canada; Environment Canada, *Guidelines for Valuers*.

15 Environment Canada, *Guidelines for Valuers*.

16 Appraisal Institute of Canada.

17 Environment Canada, *Ecological Gifts: Implementing Provisions of the Income Tax Act of Canada*, October 2000, p. 2.

A

A&B Sound, Calgary, Alberta, 195
Aberdeen Pavilion, Ottawa, Ontario, 73–75
Aberhart House, Calgary, Alberta, 169
A.E. Cross Garden Café, Calgary, Alberta, 192
Alberta, cultural resources. *See also* Calgary; Edmonton
    Father Lacombe Chapel, St. Albert, 79–80
    Fort George – Buckingham House, Elk Point, 81
    Frank Slide, Crowsnest Pass, 78–79
    Head-Smashed-In Buffalo Jump, 79
    Hillcrest Cemetery, Crowsnest Pass, 78, 79, 84
    Leitch Collieries Provincial Historic Site, Crowsnest Pass, 89
    Medalta Potteries, Medicine Hat, 2, 81
    Nordegg Mine Site, 81
    Palmer Ranch, Waterton National Park, 197
    Ralph Connor Church, Canmore, 92
    St. Patrick's Roman Catholic Church, Midnapore, 113–14
    Stephansson House, Markerville, 89
    Turner Valley Gas Plant, Turner Valley, 81
    Wayne Hotel, Wayne, 89–90
Alberta, heritage covenants, 19–20, 58, 85
Alberta, heritage preservation, 77–93. *See also* Calgary; Edmonton
    about, 1, 83, 84
    advisory services programs, 82
    of archaeological resources, 84–85
    compensation issue, 1, 5, 114–16
    funding programs, 82
    heritage designation status/process, 1, 5, 77–78, 83–84, 84
    impact assessments, 84
    intervention guidelines, 85–87
    legislation, 1, 10–11, 59, 77, 83–85, 123, 214–15
    preservation strategies, 87–93
    private trusts, 2–3
    provincial resources, 84
    registered resources, 83
    regulatory programs, 82
Alberta Heritage Resource Management Branch, 77, 82–83
Alberta Historical Resources Act, 1, 11, 59, 77, 83–85, 123, 214–15
Alberta Historical Resources Foundation, 10, 84
Alberta, preservation strategies
    adaptive re-use, 90
    alterations and additions, 92
    guidelines, 85–87
    maintenance, 89
    rehabilitation, 85–87, 90, 91–92
    restoration, 89–90
    stabilization, 89
Anderson Apartments, Calgary, Alberta, 96–97
appraisals
    "before and after" method, 124–26, 129–32, 143–44
    for cultural property *vs.* ecologically sensitive lands, 141–45
    Fair Market Value method, 135–36, 142
    in United States, 136–37
Art Gallery of Calgary, Alberta, 115–16

B

Balboa, 178–79, 179
Balmoral School, Calgary, Alberta, 100
Bankers Hall, Calgary, Alberta, 183–84, 197
Barron Building, Calgary, Alberta, 169–74
Battlefield House, Stoney Creek, Ontario, 68

Benares, Mississauga, Ontario, 65
Britain, conservation covenants
  absolute prohibitions, 26
  approvals, 27–29
  effectiveness of, 32–33
  enforcement of, 31
  interpretation of, 30
  level of control, issues with, 25
  management implications, 29–30
  planning and listed building process, 29
  relaxation of, 25–27, 27–28
  subsequent owners, issues with, 24
  and tax system, 32–33
  trends, 33
  types of, 22
Britain, cultural resources
  Covent Garden, London, 21
  Dorchester Abbey, Oxfordshire, 26–27
  Dunham Massey, Cheshire, 22, 28
  Hambleden Estate, Berkshire, 23, 33
  Zennor Head, Cornwall, 30
Britain, heritage preservation
  about, 9, 21
  legislation, 22
  and the National Trust, 20–33
British Columbia, heritage preservation, 58, 59, 60, 98
Bruce Trail, Ontario, 62–63, 63–64
building codes, 4
  Ontario Building Code Act, application of, 61–62
Burns Building, Calgary, Alberta, 176–78
Butler, Tim, 205
  The British Experience, 19–33
  concluding observations, 202
Byrne, Bill, 210
  concluding observations, 199–201
  on covenanting and development, 189–90

C

Calgary Civic Trust
  about, 2, 11–12
  conference initiative, 3, 12–14
  sample covenant and agreement, 216–35
Calgary, cultural resources
  A&B Sound, 195
  Aberhart House, 169
  A.E. Cross Garden Café, 192
  Anderson Apartments, 96–97
  Art Gallery of Calgary, 115–16
  Balmoral School, 100
  Bankers Hall, 183–84, 197
  Barron Building, 169–74
  Burns Building, 176–78
  Canada Life Building, 80
  Central High School, 116
  Central Memorial Park, 99–100
  Centre for the Performing Arts, 177–78
  Clarence Block, 116, 117
  Cliff Bungalow School, 99
  Convention Centre, 178–79, 179
  Cross House, 192
  Dr. Hughes' House, 159–61
  Grand Theatre, 110–13
  Hollingsworth Building, 80, 184
  Hyatt Hotel and Convention Centre, 104–8, 179
  Inglewood, 195, 196
  Lorraine Apartment Building, 161–65
  McDougall Cairn, 14
  Model Milk Building, 100, 101
  Mount Royal, 193
  Naismith House, 152–59
  Neilson Block, 106, 107–9
  Norman Block, 100, 103–4
  Oddfellows Temple, 114–15
  Old #1 Fire Hall, 91–92
  Palace Theatre, 101–3
  Pilkington Warehouse, 194–95
  Robertson House, 166–69
  Rouleau House, 97–98
  St. Mary's High School, 87

Stringer House, Mount Royal, 99
8th Avenue (Stephen Avenue), 193–97
Calgary, heritage preservation, 95–117
   building review process, 109–10
   density transfers, 114–16
   development process, 104
   inventory of heritage sites, 95
   municipal taxation of heritage buildings, 13
   site evaluation criteria/procedure, 96–109
Canada. Department of Canadian Heritage, 13
Canada, heritage preservation
   about, 1–3, 6–11
   Heritage Canada, 2–3, 6–7, 9, 13, 133–39
   heritage designation status, 60
   legislation, 6–7, 9, 58–61
   National Historic Sites and Monuments Board, 2, 9–10, 64
   terminology, 58, 122–23
Canada Life Building, Calgary, Alberta, 80
Canadian Customs and Revenue Agency (CCRA), 9, 13, 14, 131, 183, 184
Canadian National Railway Station, Hamilton, Ontario, 65–66
Cattle Castle, Ottawa, Ontario, 73–75
Central High School, Calgary, Alberta, 116
Central Memorial Park, Calgary, Alberta, 99–100
Centre for the Performing Arts, Calgary, Alberta, 177–78
Clarence Block, Calgary, Alberta, 116, 117
Cliff Bungalow School, Calgary, Alberta, 99
co-stewardship issues, 49–50
Collins, Jeremy, 205–6
   concluding observations, 202–3
   Ontario Precedents, 57–76
compensation issue, 1, 5–6, 60, 114–16
Constitution Act of Canada, 1–2
Convention Centre, Calgary, Alberta, 178–79, 179
Covent Garden Area Trust, Britain, 21
Cow Palace, Ottawa, Ontario, 73–75
Cross House, Calgary, Alberta, 192
Crown, The, and property rights, 4, 5–6

D

demolition
   in Alberta, 109–13
   in Ontario, 60–61
Denhez, Marc, 13, 14, 206
   on Barron Building, Calgary, 174
   concluding observations, 201–2
   on covenanting and development, 190–91, 194
   on Lorraine Apartment Building, Calgary, 164–65
   on Naismith House, Calgary, 158–59
   on Robertson House, Calgary, 168–69
   Tax Implications: Financial and Legal Considerations, 121–32
deteriorated property, issues with, 47
development industry, 7, 14
   impact of covenanting on (panel discussion), 175–98
Devonian Foundations, 2
donations, 13, 14
   Ecological Gifts Program (Canada), 142–43
   and the *Income Tax Act* of Canada, 128–29, 134–35
   in United States, 45, 136–37
Dorchester Abbey, Oxfordshire, Britain, 26–27
Douglas, Don, 14, 210
   on covenanting and development, 180–83, 184, 185–89, 194, 197
Dr. Hughes' House, Calgary, Alberta, 159–61
Drayton Hall, South Carolina, 42
Dundurn Castle, Hamilton, Ontario, 65, 66
Dunham Massey, Cheshire, Britain, 22, 28

E

Eastman Hill Stock Farm, Maine, 49–50
Edmondson, Paul, 207
- *The American Experience,* 35–54
- concluding observations, 203

Edmonton, cultural resources, 1
- Emily Murphy House, 81
- Holgate Mansion, 81
- Legislative Assembly, 89
- Lieutenant Governor of Alberta's Mansion, 89
- Rutherford House, 80

Edwards, Lucile, 152–53
Emily Murphy House, Edmonton, Alberta, 81
English Heritage, Britain, 21

F

Father Lacombe Chapel, St. Albert, Alberta, 79–80
Fleetwood Creek, Ontario, 63
Fort George – Buckingham House, Elk Point, Alberta, 81
Frank Slide, Crowsnest Pass, Alberta, 78–79
Franklin, Doug, 207–8
- *Heritage Canada Policy,* 133–39

G

Gage House, Stoney Creek, Ontario, 68
gifts, 13, 14
- Ecological Gifts Program (Canada), 142–43
- and the *Income Tax Act* of Canada, 128–29, 134–35
- in United States, 45, 136–37

Graham, Rob, 208–9
- *The Calgary Context,* 95–117
- on covenanting and development, 194–95

Grand Theatre, Calgary, Alberta, 110–13
grants, 7–8, 39–40, 84

H

Hambleden Estate, Berkshire, Britain, 23, 33
Head-Smashed-In Buffalo Jump, Alberta, 79
Heritage Canada, 2–3, 6–7, 9, 13, 133–39
heritage covenants (easements)
- about, 3, 4, 121–23
- in Alberta, 19–20
- in Britain (*See* Britain, conservation covenants)
- Calgary Civic Trust sample, 216–35
- in Canada, 6–11, 58–61
- and designation, 12–13
- explanation of terminology, 4, 35, 58, 122–23
- model development workshop (Pannekoek), 151–74
- nominal value, 127
- positive, 20
- and property rights, 123–24
- questions related to, 6
- resources list, 238–39
- restrictive, 20
- taxation issues, 8–9, 13, 35, 36, 37–39, 60, 121–32, 133–39
- value of (*See* valuation)
- web sites, 236–37

heritage designation status
- in Alberta, 1, 5, 77–78, 83–84, 84
- in Canada, 60
- in Ontario, 60–61, 62
- and "taking" of property rights, 13
- *vs.* covenants, 12–13

heritage preservation legislation
- in Alberta, 1, 10–11, 77, 83–85, 123
- in Britain, 22
- in Canada, 6–7, 9, 58–61
- in Ontario, 59, 61–62, 62
- in United States, 36–38

Heritage Railway Stations Protection Act, 1–2, 64
Hillcrest Cemetery, Crowsnest Pass, Alberta, 78, 79

Holgate Mansion, Edmonton, Alberta, 81
Hollingsworth Building, Calgary, Alberta, 80, 184
Holmes, Bob, 14, 210–11
  on covenanting and development, 175–80, 186–87, 191–94, 193–94, 197–98
Hughes' House, Calgary, Alberta, 159–61
Hyatt Hotel and Convention Centre, Calgary, Alberta, 104–8, 179

I

inalienable land, 21
Income Tax Act
  and ecologically sensitive lands, 134–35
  and gifts/donations, 128–29
Inglewood, Calgary, Alberta, 195, 196

J

Jacksonville, Oregon, 44
Jennings, Sally, on Lorraine Apartment Building, Calgary, 161–65

K

Kalman, Harold, 95
Kohler Mansion, Riverbend, Wisconsin, 48

L

lease agreements
  about, 20
  in Britain, 23, 33
  in United States, 35–36
Legislative Assembly, Edmonton, Alberta, 89
Leitch Collieries Provincial Historic Site, Crowsnest Pass, Alberta, 89
Lieutenant Governor of Alberta's Mansion, Edmonton, Alberta, 89
Lorraine Apartment Building, Calgary, Alberta, 161–65

Lowell's Boat Shop, Massachusetts, 46, 47
Lyndhurst, New York, 42–43

M

MacLaren, Fergus, on Robertson House, Calgary, 167–68
maintenance of property, issues with, 48
Manitoba, heritage preservation, 59
Mar a Lago, Palm Beach, Florida, 50–53
Mattapoisett, Massachusetts, 45–46
McDougall Cairn, Calgary, Alberta, 14
McMordie, Michael, 177, 211–12
Medalta Potteries, Medicine Hat, Alberta, 2, 81
Milavsky, Harold, 14, 212
  on covenanting and development, 183–85, 187, 193, 197
Model Milk Building, Calgary, Alberta, 100, 101
Montpelier in Orange, Virginia, 43
Mount Royal, Calgary, Alberta, 193
Myrtle Grove, Maryland, 53

N

Naismith House, Calgary, Alberta, 152–59
National Historic Sites and Monuments Board of Canada, 2, 9–10, 64
National Trust, Britain
  about, 20–21
  and conservation covenants, 22–33
National Trust Act, Britain, 22
National Trust for Historic Preservation, United States, 35
natural heritage easements, 59
Neilson Block, Calgary, Alberta, 106, 107–9
Ness, Jason, 209
  Calculating the Market Value: Appraisal Methods, 141–45
  Naismith House discussion, 153–59
New Baltimore, New York, 47
New Brunswick, heritage preservation, 59

Newfoundland, heritage preservation, 59, 60
Niagara Apothecary, Niagara-on-the-Lake, Ontario, 63
Nordegg Mine Site, Alberta, 81
Norman Block, Calgary, Alberta, 100, 103–4
Northwest Territories, heritage preservation, 59
not-for-profit organizations, 2, 3
Nova Scotia, heritage preservation, 59
Nunavut, heritage preservation, 59

O

Oatlands, Virginia, 41
Oddfellows Temple, Calgary, Alberta, 114–15
Old #1 Fire Hall, Calgary, Alberta, 91–92
Old Strathcona Foundation, Alberta, 2
Ontario, cultural resources
   Aberdeen Pavilion, Ottawa, 73–75
   Battlefield House, Stoney Creek, 68
   Benares, Mississauga, 65
   Bruce Trail, 62–63, 63–64
   Canadian National Railway Station, Hamilton, 66, 67
   Dundurn Castle, Hamilton, 65, 66
   Fleetwood Creek, 63
   Gage House, Stoney Creek, 68
   Niagara Apothecary, Niagara-on-the-Lake, 63
   Ruthven Park Estate, Cayuga, 67–68
   St. Brigid's Roman Catholic Church, Ottawa, 65–66
Ontario, heritage easements (covenants), 3, 59, 63–76
   acquisition of, 64–65, 70–71
   alteration requests and approvals, 73
   baseline documentation report, 71–72
   contents of, 69–70
   defined, 57–58
   monitoring and enforcement, 72
   ownership transition information package, 72–73
   publicizing, 72
   tailoring easement to property, 71
Ontario, heritage preservation
   court case, St. Michael's Church, Cobourg, 61–62
   and demolition control, 60–61
   and heritage designation status, 60–61, 62
   of historic battlegrounds, 68
   of historic fairground carousel, 68, 69
   legislation, 59, 61–62, 62
   of natural heritage sites, 62
   Ontario Heritage Foundation, 3, 59, 62–76
   portfolio of heritage sites, 62–63, 65–69
   religious easement properties, 65–66

P

Palace Theatre, Calgary, Alberta, 101–3
Palmer Ranch, Waterton National Park, Alberta, 197
Pannekoek, Frits, 151, 212–13
Parks Canada, 9
Pearson, Larry, 209
Preservation and Planning Context in Alberta, 77–93
PetroCanada, Calgary, Alberta, 114
Pilkington Warehouse, Calgary, Alberta, 194–95
positive covenants, 20
Post, Marjorie Merriweather, 50
preservation advocates, 7
preservation strategies. *See* Alberta, preservation strategies
Prince Edward Island, heritage preservation, 58, 59
private trusts, 2–3
property rights, 1–2, 4–6, 123–24, 141
Provincial Historic Resources (Alberta), 7–8
public access issue, in Ontario, 70

Q

Quebec, heritage preservation, 58, 59

R

railway stations, protection of, 1–2, 10, 64–65, 67
Ralph Connor Church, Canmore, Alberta, 92
resources list, for heritage covenants (easements) information, 238–39
restrictive covenants, 20
Robertson, Allison, 159
Robertson, Nora, 166–67
    on covenanting and development, 195–96
Robertson House, Calgary, Alberta, 166–69
Roman Catholic Episcopal Corporation for the Diocese of Peterborough v. Corporation of the Town of Cobourg, 61–62
Rouleau House, Calgary, Alberta, 97–98
Rutherford House, Edmonton, Alberta, 80
Ruthven Park Estate, Cayuga, Ontario, 67–68

S

Saskatchewan, heritage preservation, 59
servitudes, 58
Shaw, Fraser, on Barron Building, Calgary, 169–74
SMED, 195
Smith, Kelly, 159
St. Brigid's Roman Catholic Church, Ottawa, Ontario, 65–66
St. Mary's High School, Calgary, Alberta, 87
St. Patrick's Roman Catholic Church, Midnapore, Alberta, 113–14
Stephansson House, Markerville, Alberta, 89
Stephen Avenue (8th Avenue), Calgary, Alberta, 193–97
Stringer House, Mount Royal, Calgary, Alberta, 99

T

taxation issues, 8–9, 13, 35, 36, 37–39, 60, 121–32, 133–39
terminology, 4
    in Canada, 58, 122–23
    in United States, 35
8th Avenue (Stephen Avenue), Calgary, Alberta, 193–97
The Mount, Lenox, Massachusetts, 48–49
Todd House, West Virginia, 48
Troth's Fortune, Maryland, 44, 45
Trump, Donald, 50–52
Turner Valley Gas Plant, Turner Valley, Alberta, 81

U

United States, cultural resources
    Drayton Hall, South Carolina, 42
    Eastman Hill Stock Farm, Maine, 49–50
    Jacksonville, Oregon, 44
    Kohler Mansion, Riverbend, Wisconsin, 48
    Lowell's Boat Shop, Massachusetts, 46, 47
    Lyndhurst, New York, 42–43
    Mar a Lago, Palm Beach, Florida, 50–53
    Mattapoisett, Massachusetts, 45–46
    Montpelier in Orange, Virginia, 43
    The Mount, Lenox, Massachusetts, 48–49
    Myrtle Grove, Maryland, 53
    New Baltimore, New York, 47
    Oatlands, Virginia, 41
    Todd House, West Virginia, 48
    Troth's Fortune, Maryland, 44, 45
United States, heritage easements (covenants), 9, 35–54
    amending, 53
    British system comparison, 35–36
    costs, 40–41
    covenant *vs.* easement, 35
    drafting, 52
    economic benefits, 35, 39–40

and endowments, 52
lease arrangements, 35–36
legal basis of, 36–38
National Trust for Historic Preservation, 36, 41–54
tax rules and benefits, 37–39
valuation/appraisals, 136–37
and zoning issues, 52

V

valuation, 4–5, 135–36
"before and after" method, 124–26, 129–32
for cultural property *vs.* ecologically sensitive lands, 141–45
Fair Market Value method, 135–36, 142
in United States, 136–37
Vancouver, heritage preservation, 98
von Kuster, D.W., Dr. Hughes House, Calgary, 160–61

W

Wayne Hotel, Wayne, Alberta, 89–90
web sites, for heritage covenants (easements) information, 236–37
Wharton, Edith, 48

Y

Yukon, heritage preservation, 59

Z

Zennor Head, Cornwall, Britain, 30
zoning restrictions, 4–5, 52